# Introduction to the Theories of Measurement and Meaningfulness and the Use of Symmetry in Science

# THE LEA SERIES IN SCIENTIFIC PSYCHOLOGY
*Edited by Stephen W. Link, University of California, San Diego and James T. Townsend, Indiana University*

*Introduction to the Theories of Measurement and Meaningfulness and the Use of Symmetry in Science* by Louis Narens

*Measurement and Representation of Sensations* edited by Hans Colonius and Ehtibar N. Dzhafarov

*Psychophysics Beyond Sensation* edited by Christian Kaernbach, Erich Schroger, and Hermann Muller

*Theories of Meaningfulness* by Louis Narens

*Empirical Direction in Design and Analysis* by Norman H. Anderson

*Computational, Geometric, and Process Perspectives on Facial Cognition* edited by Michael J. Wenger and James T. Townsend

*Utility of Gains and Losses* by R. Duncan Luce

*The War Between Mentalism and Behaviorism* by William R. Uttal

*Recent Progress in Mathematical Psychology* edited by Cornelia E. Dowling, Fred S. Roberts, and Peter Theuns

*Localist Connectionist Approaches To Human Cognition* edited by Jonathan Grainger and Arthur M. Jacobs

*Sensation and Judgment* by John Baird

*Adaptive Spatial Alignment* by Gordon M. Redding and Benjamin Wallace

*Signal Detection Theory and Roc Analysis in Psychology and Diagnostics* by John Swets

*The Wave Theory of Difference and Similarity* by Stephen W. Link

*The Swimmer* by William R. Uttal, Gary Bradshaw, Siram Dayanand, Robb Lovell, Thomas Shepherd, Ramakrishna kakarala, Kurt Skifsted, and Greg Tupper

*Multidimensional Models of Perception and Cognition* by Gregory F. Ashby

*Cognition, Information Processing, and Psychophysics* edited by Hans-Georg Geissler, Stephen W. Link, and James T. Townsend

# Introduction to the Theories of Measurement and Meaningfulness and the Use of Symmetry in Science

Louis Narens
*University of California, Irvine*

LAWRENCE ERLBAUM ASSOCIATES, PUBLISHERS
2007    Mahwah, New Jersey                London

Camera ready copy for this book was provided by the author.

Copyright © 2007 by Lawrence Erlbaum Associates, Inc.

All rights reserved. No part of this book may be reproduced in any form, by photostat, microform, retrieval system, or any other means, without prior written permission of the publisher.

Lawrence Erlbaum Associates, Inc., Publishers
10 Industrial Avenue
Mahwah, New Jersey 07430
www.erlbaum.com

Cover design by Tomai Maridou

**Library of Congress Cataloging-in-Publication Data**

Narens, Louis.
Introduction to the theories of measurement and meaningfulness and the use of symmetry in science / Louis Narens.
     p.  cm.
Includes bibliographical references and index.
ISBN 978-0-8058-6202-7 — 0-8058-6202-1 (cloth)
ISBN 978-1-4106-1598-5 — 1-4106-1598-7 (e book)
1. Mensuration. 2. Symmetry (Mathematics) 3. Threshold logic. I. Title.
QA465.N374  2006
530.801—dc22                                  2006022396
                                                                                                      CIP

Books published by Lawrence Erlbaum Associates are printed on acid-free paper, and their bindings are chosen for strength and durability.

Printed in the United States of America
10  9  8  7  6  5  4  3  2  1

*For Duncan Luce, whose visionary ideas on measurement and psychophysics inspired this volume*

# Contents

| | | |
|---|---|---|
| **1** | **The Need for Theories of Measurement and Meaningfulness** | **1** |
| | 1.1 Plateau's Theory | 2 |
| | 1.2 Derivation of Plateau's Power Law | 4 |
| | 1.3 Issues Raised by Plateau's Theory | 8 |
| | 1.4 Additional Lemmas and Proofs | 10 |

| | | |
|---|---|---|
| **I** | **Measurement** | **15** |
| **2** | **Representational Measurement Theory** | **17** |
| | 2.1 Introduction | 17 |
| | 2.2 Preliminaries | 18 |
| |     2.2.1 Basic Notation | 18 |
| |     2.2.2 Isomorphisms | 18 |
| | 2.3 Cantor's Characterization of the Continuum | 20 |
| | 2.4 Continuous Structures | 22 |
| | 2.5 Scale Types | 28 |
| | 2.6 Representational Meaningfulness | 32 |
| **3** | **Symmetries and the Erlanger Program** | **35** |
| | 3.1 Symmetries | 35 |
| | 3.2 Erlanger Program | 40 |
| | 3.3 Comparison of Geometric and Measurement-Theoretic Concepts | 43 |
| **4** | **Threshold Measurement** | **47** |
| | 4.1 Continuous Threshold Structures | 48 |
| | 4.2 Weber's and Fechner's Laws | 52 |
| | 4.3 Threshold Structures with Only Psychological Primitives | 56 |
| | 4.4 Meaningfulness Considerations | 57 |

## 5 Magnitude Production — 61
- 5.1 Stevens' Methods of Magnitude Estimation and Production — 61
- 5.2 Narens' 1996 Theory — 63
- 5.3 Empirical Tests — 68
- 5.4 Continuous Ratio Production — 69
- 5.5 Conclusions — 70

## 6 Torgerson's Conjecture — 73
- 6.1 Bisection Data — 73
- 6.2 Torgerson's Conjecture — 77
- 6.3 An Experimental Test of Torgerson's Conjecture — 79
- 6.4 Theoretical Considerations — 81
- 6.5 Proof of Lemma 6.1 — 82

# II Meaningfulness — 87

## 7 Meaningfulness Concepts from Measurement Theory — 89
- 7.1 Quantitative $\mathcal{S}$-Meaningfulness — 89
- 7.2 Qualitative $\mathcal{S}$-Meaningfulness — 95
- 7.3 Endomorphism Invariance — 97

## 8 Preliminary Set Theory — 101
- 8.1 Introduction — 101
- 8.2 The Language $\mathsf{L}(\in, \boldsymbol{A})$ — 101
- 8.3 Basic Set Theory — 105
- 8.4 The Sets $\mathsf{V}$ and $\mathsf{P}$ — 109
- 8.5 First-Order and Higher-Order Relations — 112

## 9 Scientific Topics — 115
- 9.1 Principles for Scientific Topics — 115
- 9.2 Pure Mathematics — 117
- 9.3 Set-Theoretic Interpretation — 118

## 10 Theories of Meaningfulness — 121
- 10.1 Axiom of Measurement — 121
- 10.2 Axiom System $\mathcal{FST}$ — 126
- 10.3 Meaningfulness and the Erlanger Program — 127
  - 10.3.1 Invariance under Extensions of Permutations — 127
  - 10.3.2 $\in$-Symmetries — 129
  - 10.3.3 Axiom System $\mathcal{E}$ — 131
- 10.4 Additional Proofs — 135

## 11 Applications, Limitations, and Generalizations of Axiom System $\mathcal{FST}$ — **143**
- 11.1 An Epistemology for a Rule Based on Invariance — 143
- 11.2 Limitations of Axiom System $\mathcal{FST}$ — 150
- 11.3 Intrinsicness — 151
- 11.4 Possible Psychophysical Laws — 155
- 11.5 Distinguishing Empirical and Meaningful Relations — 161

## References — **163**

## Index — **169**

## Chapter 1

# The Need for Theories of Measurement and Meaningfulness

Measurement brings numbers into science. More formally, *measurement* is the assignment of numbers or mathematical objects to empirical or qualitative entities. The way a sub-area of science accomplishes measurement impacts the form of its mathematical representation. Mathematical science routinely applies powerful methods of mathematical inference to the representation to infer relationships among the measured objects. The problem of sorting out which of these relationships belong to the sub-area of science under consideration and which are "just mathematical" or perhaps belong to a different sub-area of science is called the *meaningfulness problem*. Although measurement and meaningfulness are considered important in various parts sciences, their presentations in the literature are usually informal and superficial. This is somewhat surprising given the numerous debates in the scientific literature about what can and cannot be concluded from specific mathematical models and data sets. Considerations involving measurement and meaningfulness are also essential in providing a proper foundation for various techniques in science that make inferences based on invariance and symmetry.

In the philosophy of science, the role of mathematics in science is often omitted or discussed in a discursive manner that avoids many of the subtle foundational issues concerning measurement, meaningfulness, and the uses of invariance and symmetry—issues that can only be made apparent and analyzed in a well-developed, formal mathematical theory.

This book presents such a theory. Its measurement part is a variant of the Representational Theory of Measurement (Pfanzagl, 1968; Krantz, Luce, Suppes, & Tversky, 1971; Suppes, Krantz, Luce, & Tversky, 1990; Luce, Krantz, Suppes, & Tversky, 1990; Narens, 1985), the current dominant theory of measurement. Its meaningfulness part is a shortened version of one of the theories of meaningfulness presented in Narens (2002a), and its invariance and symmetry parts come from theory and examples developed in Narens (2002a).

The book is designed to be an introduction to the theories of measurement and meaningfulness and not a comprehensive study of those topics. To keep it short, many of the standard theorems of measurement theory are only stated with references to the literature for proofs, and only one theory of meaningfulness is considered in detail. The mathematics used is essentially self-contained. The volume is at the level of an upper division mathematics course at an American university. However, the ability to think mathematically and abstractly about scientific and philosophical topics is also needed.

The examples in the book are mostly from an area of psychology known as *psychophysics*. Psychophysics studies how physical properties are neurally encoded and represented as mental phenomena. It is particularly interesting from the standpoint of measurement and meaningfulness, because it involves the measurements of both physical and mental phenomena and mathematical models relating them.

A major theme of this book is the psychophysical measurement of subjective intensity. This has been a subject of intense interest in psychology from the very beginning of experimental psychology. And from that beginning to the present day, it has continuously generated major controversies involving measurement and meaningfulness.

## 1.1 PLATEAU'S THEORY

Fechner founded experimental psychology with the publication of his *Elemente der Psychophysik (Elements of Psychophysics)* (Fechner, 1860). The crowning achievement of this work is a theory and method for measuring the subjective intensities of physical stimuli, for example, the subjective brightness of light. Fechner's method is an example of what today is called *indirect measurement*. The term "indirect" is used to indicate that the measurement is based on objective observations of behavior, rather than, for example, "direct" subjective estimates of intensity. A modern measurement-theoretic approach to Fechner's method is presented in chapter 4.

Fechner was interested in determining the shape of the *psychophysical function*—that is, the shape of the function that maps the measurements of stimuli from a physical continuum, for example, the physical intensity of lights, onto a numerical continuum measuring an observer's perceived intensities of the stimuli, for example, the perceived brightness of lights. Using his theory of measurement combined with extensive empirical observations, Fechner concluded that the psychophysical function was logarithmic.

In contrast, Plateau (1872) concluded that the psychophysical function was a power function, that is, a function of the form $f(r) = ar^b$, where $a$ and $b$ are positive constants and $r$ is a positive variable. His argument used mathematical reasoning based on invariance considerations and a few empirical instances involving of what he interpreted to be subjective judgments of ratios of the brightnesses of stimuli. Such judgments are examples of what today are called *direct measurements*. The term "direct" is used to indicate that they are based on the assignment of numerical relationships that veridically match the observer's verbally based evaluations of relationships of perceived intensities.

Plateau's empirical work consisted of providing eight artists with two disks—one painted black and the other white—with the instruction to paint a gray disk midway between them. He reported that the resulting eight gray disks were almost identical, despite differences in illumination under which they were painted. He assumed each artist mixed his gray paint to obtain a gray such that the ratio of the subjective intensity of white to gray equaled the subjective intensity of gray to black. Assuming a similar result for any pair of gray disks (where throughout this section, by convention, black and white disks are considered extreme examples of gray disks), then yields,

$$\frac{\psi(d)}{\psi(m)} = \frac{\psi(m)}{\psi(e)},$$

where $d$ and $e$ are the disks provided for midway judgment, $m$ is the midway disk painted by the artist, and $\psi$ is a function that measures the subjective intensity of grayness. It is a known physical fact that the *ratios* of the usual physical measurements of the light of such gray disks do not vary with illumination. Thus, in particular such ratios would be the same in each artist's studio. Using this observation, Plateau concluded that his experiment exhibited the following law for his stimuli.

**Definition 1.1 (Preserved Midway Ratio Law)** The *Preserved Midway Ratio Law* holds if and only if or all gray disks $d$ and $e$,

$$\frac{\varphi(d)}{\varphi(m)} = \frac{\varphi(m)}{\varphi(e)} \text{ iff } \frac{\psi(d)}{\psi(m)} = \frac{\psi(m)}{\psi(e)},$$

where $m$ is the gray disk produced midway between $d$ and $e$, $\varphi$ is a function that measures the physical intensity of grays, and $\psi$ is a function that measures subjective grayness. □

Using the the Preserved Midway Ratio Law, it follows mathematically that the psychophysical power law holds, that is,

$$\psi = r\varphi^s$$

for some positive $r$ and $s$; that is, the psychophysical function is a power function. (This is shown in Theorem 1.2.)

## 1.2 DERIVATION OF PLATEAU'S POWER LAW

Because psychophysical functions play a central role in this book, it is useful for future reference to have the following formal definition of them.

**Definition 1.2 (Psychophysical Function)** Suppose $X$ is a set of physical stimuli measured by the function $\varphi$ onto the real numbers. (Generally, throughout this book such a $\varphi$ will be taken to be one of the usual standard measuring functions use in physics. It is worthwhile to note that such measuring functions are objective in the sense that different experimenters could independently construct the same measuring functions for the same stimuli.) Suppose $\psi$ is a function on $X$ into the real numbers that measures the subjective intensity of the elements of $X$. ($\psi$ is subjective and generally is not observable to the experimenter. The non-objectivity and non-observability of $\psi$ pose interesting and challenging methodological and philosophical issues for the scientific study of subjective intensity.) Let $\Psi$ be the function such that for all $x$ in $X$,

$$\psi(x) = \Psi(\varphi(x))\,.$$

$\Psi$ is called the *psychophysical function* (on $X$ relating $\varphi$ and $\psi$), $\varphi$ is called the *physical function* (measuring $X$), and $\psi$ is called *psychological function* (measuring $X$). $\Psi$ is said to be *logarithmic* if and only if for some real $c$ and positive $d$, $\Psi(r) = d \cdot \log(r) + c$, and $\Psi$ is said to be a *power function* if and only if for some positive $a$ and $b$, $\Psi(r) = a \cdot r^b$. (The constants $c$ and $a$ depend on the unit of the physical function measuring $X$. Thus measuring $X$ by the physical function $t \cdot \varphi$ changes $c$ into $c + d \cdot \log(t)$ and $a$ into $a \cdot t^b$.) □

**Convention 1.1** Throughout this book, $\mathbb{R}$ stands for the set of real numbers and $\mathbb{R}^+$ for the set of positive real numbers. □

The derivation of Plateau's Power Law from the Preserved Midway Ratio Law presented in this section uses the following lemma from mathematics. The lemma involves the solution of a functional equation. The last section of this chapter provides a proof for this lemma as well as the proofs for a few other lemmas involving functional equations. These lemmas are used in later chapters. Their proofs only involve elementary properties of the real numbers and simple algebraic techniques.

**Lemma 1.1** *Suppose $\Psi$ and $K$ are strictly increasing functions from $\mathbb{R}^+$ onto $\mathbb{R}^+$ such that for all $\beta$ and $r$ in $\mathbb{R}^1$,*

$$\Psi(\beta r) = K(\beta)\Psi(r). \tag{1.1}$$

*Then for some positive $\lambda$ and $\gamma$, $\Psi(r) = \lambda r^\gamma$.*
**Proof.** Lemma 1.6. □

The following theorem is key to the derivation presented in this chapter of Plateau's Power Law.

**Theorem 1.1** *Suppose $X$ is a set of physical stimuli, $\varphi$ is a physical measuring function from $X$ onto the positive reals, $\psi$ is a psychological measuring function from $X$ onto the positive reals, and $\varphi$ and $\psi$ are strictly monotonically related, that is, for all $x$ and $y$ in $X$,*

$$\varphi(x) < \varphi(y) \text{ iff } \psi(x) < \psi(y). \tag{1.2}$$

*Then the following two statements are equivalent:*

1. **Psychophysical Power Law**: *There exists positive real numbers $\lambda$ and $\gamma$ such that for all $x$ in $X$,*

$$\psi(x) = \lambda\varphi(x)^\gamma.$$

2. **Preserved Equal Ratios Law**: *For all $x$, $y$, $u$, and $v$ in $X$,*

$$\frac{\varphi(x)}{\varphi(y)} = \frac{\varphi(u)}{\varphi(v)} \text{ iff } \frac{\psi(x)}{\psi(y)} = \frac{\psi(u)}{\psi(v)}. \tag{1.3}$$

**Proof.** Suppose the Psychophysical Power Law, that is, suppose $\lambda$ and $\gamma$ are positive reals such that for all $x$ in $X$,

$$\psi(x) = \lambda\varphi(x)^\gamma.$$

To show the Preserved Equal Ratios Law, let $x$, $y$, $u$, and $v$ be arbitrary elements of $X$. Then

$$\frac{\varphi(x)}{\varphi(y)} = \frac{\varphi(u)}{\varphi(v)} \text{ iff } \left(\frac{\varphi(x)}{\varphi(y)}\right)^\gamma = \left(\frac{\varphi(u)}{\varphi(v)}\right)^\gamma$$

$$\text{iff } \frac{\lambda\varphi(x)^\gamma}{\lambda\varphi(y)^\gamma} = \frac{\lambda\varphi(u)^\gamma}{\lambda\varphi(v)^\gamma}$$

$$\text{iff } \frac{\psi(x)}{\psi(y)} = \frac{\psi(u)}{\psi(v)}.$$

Suppose the Preserved Equal Ratios Law. Let $x$, $y$, $u$, and $v$ be arbitrary elements of $X$ such that
$$\frac{\varphi(x)}{\varphi(y)} = \frac{\varphi(u)}{\varphi(v)} \text{ iff } \frac{\psi(x)}{\psi(y)} = \frac{\psi(u)}{\psi(v)}. \tag{1.4}$$
Defined $\Psi$ on $\mathbb{R}^+$ as follows: For all $x$ in $X$,
$$\Psi(\varphi(x)) = \psi(x).$$
It then follows from Equation 1.2 and the definitions of $\varphi$ and $\psi$ that $\Psi$ is a strictly increasing function from $\mathbb{R}^+$ onto $\mathbb{R}^+$. Let $\beta$ and $r$ be arbitrary positive reals. Because by hypothesis $\varphi$ is onto $\mathbb{R}^+$, let $a$, $b$, $c$, and $d$ be elements of $X$ such that
$$r = \varphi(a), \ \beta = \varphi(b), \ 1 = \varphi(c), \text{ and } \beta r = \varphi(d).$$
Because
$$\frac{r}{1} = \frac{\beta r}{\beta},$$
it follows that
$$\frac{\varphi(a)}{\varphi(c)} = \frac{\varphi(d)}{\varphi(b)}.$$
Thus by Equation 1.4,
$$\frac{\psi(a)}{\psi(c)} = \frac{\psi(d)}{\psi(b)},$$
and therefore by the definition $\Psi$,
$$\frac{\Psi(\varphi(a))}{\Psi(\varphi(c))} = \frac{\Psi(\varphi(d))}{\Psi(\varphi(b))},$$
that is,
$$\frac{\Psi(r)}{\Psi(1)} = \frac{\Psi(\beta r)}{\Psi(\beta)}. \tag{1.5}$$
Letting
$$K(\beta) = \frac{\Psi(\beta)}{\Psi(1)},$$
Equation 1.5 becomes
$$K(\beta)\Psi(r) = \Psi(\beta r). \tag{1.6}$$
By Lemma 1.1,
$$\Psi(r) = \lambda r^\gamma,$$
for some positive $\lambda$ and $\gamma$. Thus, by the definition of $\Psi$, for all $z$ in $X$,
$$\psi(z) = \lambda \varphi(z)^\gamma$$
for some positive $\lambda$ and $\gamma$. □

# Theories of Measurement and Meaningfulness

**Lemma 1.2** *The Preserved Midway Ratio Law implies the Preserved Equal Ratios Law.*

**Proof.** Suppose the Preserved Midway Ratio Law. Let $X$ be a set of physical stimuli, $\varphi$ be a physical measuring function from $X$ onto the positive reals, $\psi$ be a psychological measuring function from $X$ onto the positive reals, and $\varphi$ and $\psi$ be strictly monotonically related, that is, for all $x$ and $y$ in $X$,

$$\varphi(x) < \varphi(y) \text{ iff } \psi(x) < \psi(y). \tag{1.7}$$

Let $x$, $y$, $u$, and $v$ be arbitrary elements of $X$. It needs only to be shown that

$$\frac{\varphi(x)}{\varphi(y)} = \frac{\varphi(u)}{\varphi(v)} \text{ iff } \frac{\psi(x)}{\psi(y)} = \frac{\psi(u)}{\psi(v)}.$$

*Part 1.* Suppose

$$\frac{\varphi(x)}{\varphi(y)} = \frac{\varphi(u)}{\varphi(v)}. \tag{1.8}$$

Because $\varphi$ and $\psi$ are onto the positive reals, let $p$ and $q$ in $X$ be such that

$$\frac{\varphi(x)}{\varphi(p)} = \frac{\varphi(p)}{\varphi(v)} \text{ and } \frac{\varphi(u)}{\varphi(q)} = \frac{\varphi(q)}{\varphi(y)}.$$

Then by the Preserved Midway Ratio Law,

$$\frac{\psi(x)}{\psi(p)} = \frac{\psi(p)}{\psi(v)} \text{ and } \frac{\psi(u)}{\psi(q)} = \frac{\psi(q)}{\psi(y)}.$$

Thus

$$\varphi(x)\varphi(v) = \varphi(p)^2 \text{ and } \varphi(u)\varphi(y) = \varphi(q)^2, \tag{1.9}$$

and

$$\psi(x)\psi(v) = \psi(p)^2 \text{ and } \psi(u)\psi(y) = \psi(q)^2. \tag{1.10}$$

Because, by Equation 1.8, $\varphi(x)\varphi(v) = \varphi(u)\varphi(y)$, it follows from Equation 1.9 that $\varphi(p) = \varphi(q)$. By Equation 1.7, $\varphi$ is a one-to-one function. Therefore $p = q$. Thus by $p = q$ and Equation 1.10,

$$\psi(x)\psi(v) = \psi(p)^2 = \psi(q)^2 = \psi(u)\psi(y),$$

and therefore

$$\frac{\psi(x)}{\psi(y)} = \frac{\psi(u)}{\psi(v)}.$$

*Part 2.* Suppose $\psi(x)/\psi(y) = \psi(u)/\psi(v)$. Then $\varphi(x)/\varphi(y) = \varphi(u)/\varphi(v)$ follows by a similar argument. □

**Theorem 1.2 (Plateau's Power Law)** *Suppose $X$ is a set of physical stimuli, $\varphi$ is a physical measuring function from $X$ onto the positive reals, $\psi$ is a psychological measuring function from $X$ onto the positive reals, and $\varphi$ and $\psi$ are strictly monotonically related, that is, for all $x$ and $y$ in $X$,*

$$\varphi(x) < \varphi(y) \text{ iff } \psi(x) < \psi(y).$$

*Then the following two statements are equivalent:*

1. **Psychophysical Power Law**: *There exists positive real numbers $\lambda$ and $\gamma$ such that for all $x$ in $X$,*

$$\psi(x) = \lambda \varphi(x)^\gamma.$$

2. **Preserved Midway Ratio Law**: *For all $x$, $y$, and $m$ in $X$,*

$$\frac{\varphi(x)}{\varphi(m)} = \frac{\varphi(m)}{\varphi(y)} \text{ iff } \frac{\psi(x)}{\psi(m)} = \frac{\psi(m)}{\psi(y)}.$$

**Proof.** It is immediate that the Preserved Equal Ratios Law implies the Preserved Midway Ratio Law. Thus by Lemma 1.2, the Preserved Midway Ratio Law is logically equivalent to the Preserved Equal Ratios Law. The theorem then follows from Theorem 1.1. □

## 1.3 ISSUES RAISED BY PLATEAU'S THEORY

Plateau's argument raises a number of philosophical and methodological issues. Some of these are general ones that apply to a wide range of applied mathematical arguments; others are related to general issues involving the mathematical modeling of subjective experience; still others concern the interpretation of the specific experiment carried out by Plateau.

The first two issues about the existence of measurement functions. These are addressed mainly in Part I of this book.

1. Plateau's physical measuring function $\varphi$ came from physics:

   - What assumptions are being made about the existence of such functions? To what extent are they based on empirical considerations?

   - Why should changes in illumination leave the ratios of measurements of such functions invariant? (An important assumption of Plateau.) Is it by convention? Theory? Or is there some empirical reason for this?

2. Plateau theorizes the existence of the psychological measuring function $\psi$.

- Is it reasonable that mental phenomena are capable of such measurement? To what extent can the direct measurement part of Plateau's argument be replaced by indirect methods of measurement?

3. Plateau made inferences based on symmetry and invariance, for example, deriving the Psychophysical Power Law from the Preserved Midway Ratio Law. These kinds of inferences are common in physics and are powerful and important tools in science. However, at the foundational level, there has been little formal analysis of them and little written about their proper use. (Inferences based on symmetry and invariance are used throughout the book and are often linked to measurement issues. A theory justifying such inferences is given in Part II of this book, particularly in chapter 11.)

4. Plateau's argument for a power law has two major problems involving the use and interpretation of "midway." The first is that specific testable properties of "midway" could fail empirically:

- "Midway" can be characterized mathematically as an operation on two variables: $M(x,z) = y$ if and only if $y$ is midway between $x$ and $z$. When so characterized, "midway" has certain special algebraic properties, for example,

$$M[M(x,y), M(u,v)] = M[M(x,u), M(y,v)].$$

(A complete description of these properties is given in Definition 2.14 of chapter 2.) Plateau did not check whether the artists' midway grays had the appropriate testable mathematical structural properties of a midway operation. If these structural properties were to fail empirically, then his entire argument falls apart.

- The second problem with Plateau's argument, as given in Section 1.1, is that even if his empirical midway operation satisfied all the testable algebraic properties of a "midway" operation, then there would still be no objective evidence to argue that the observer is using the midway operation to produce subjectively equal ratios, as oppose, for example, subjectively equal differences.[1] Chapter 6 presents a detailed discussion of this issue.

---

[1] Plateau was apparently aware of this possibility: Falmagne (1985) writes,

In a footnote in his [Plateau's] paper, we read "Fechner's formula leads to this consequence that, when the overall illumination increases, the differences in sensation remain constant; it seemed to me more rational, in order to explain the invariance of the general effect of the picture, to postulate *a priori* the

Variants of the last two problems appear routinely in mathematical science. One of the objectives of theories of measurement is understand how, and to what extent, they can be theoretically and practically solved.

Additional foundational issues are created by other scientific applications, for example, the psychologist S. S. Stevens' approach to psychophysical functions, or the use of dimensional analysis in physics. These and several others are discussed throughout this book.

## 1.4  ADDITIONAL LEMMAS AND PROOFS

**Lemma 1.3** *Suppose $f$ is a strictly increasing function from $\mathbb{R}^+$ onto $\mathbb{R}^+$ such that for all $x$ and $y$ in $\mathbb{R}^+$,*

$$f(x+y) = f(x) + f(y).$$

*Then for some positive real number $c$, $f(x) = c \cdot x$.*

**Proof.** Suppose $f$ is a strictly increasing function such that for all $x$ and $y$ in $\mathbb{R}^+$,

$$f(x+y) = f(x) + f(y).$$

Let $t$ be an element of $\mathbb{R}^+$ and

$$c = \frac{f(t)}{t}.$$

It will be shown by contradiction that for all $x$ in $\mathbb{R}^+$, $c \cdot x = f(x)$. Suppose $y$ in $\mathbb{R}^+$ is such that $c \cdot y \neq f(y)$. Then either $c \cdot y < f(y)$ or $c \cdot y > f(y)$. Without loss of generality suppose $c \cdot y < f(y)$. (The case for $c \cdot y > f(y)$ has an almost identical argument.) By elementary algebraic properties of the real numbers, let $m$ and $n$ be positive integers such that

$$cy < \frac{m}{n} ct < f(y).$$

Simplifying and noting that $mct = mf(t)$, we obtain

$$ncy < mct = mf(t) < nf(y). \tag{1.11}$$

From $ncy < mct$ in Equation 1.11, we obtain,

$$ny < mt. \tag{1.12}$$

---

constancy of the ratios and not the differences of the sensations." *(pg. 318)* [Translation by Falmagne from pp. 382–383 of Plateau (1872); the emphasis is Falmagne's.]

Note that $f(1x) = f(x)$, $f(2x) = f(x+x) = f(x) + f(x) = 2f(x)$, $f(3x) = f(2x + x) = f(2x) + f(x) = 2f(x) + f(x) = 3f(x)$, and by using a simple mathematical inductive argument, $f(px) = pf(x)$ for all positive integers $p$ and all positive reals $x$. Noting that $mf(t) = f(mt)$, $nf(y) = f(ny)$, and by assumption $f$ is strictly monotonic, we obtain from $mf(t) < nf(y)$ in Equation 1.11 that $f(mt) < f(ny)$, and thus that

$$mt < ny. \qquad (1.13)$$

Equations 1.12 and 1.13 produce a contradiction. □

**Lemma 1.4** *Suppose $f$ is a strictly increasing function from $\mathbb{R}$ onto $\mathbb{R}$ such that for all $x$ and $y$ in $\mathbb{R}$,*

$$f(x + y) = f(x) + f(y).$$

*Then for some positive real number $c$, $f(x) = c \cdot x$.*

**Proof.** For $x = y = 0$,

$$f(0 + 0) = f(0) + f(0) = 2f(0),$$

and thus

$$f(0) = 0.$$

Let $f^+$ be $f$ restricted to $\mathbb{R}^+$. Then for $0 < x$,

$$0 = f(0) < f(x) = f^+(x).$$

Therefore, because $f$ is strictly increasing and onto $\mathbb{R}$, $f^+$ is a strictly increasing function from $\mathbb{R}^+$ onto $\mathbb{R}^+$. Because $f^+$ is the restriction of $f$ to $\mathbb{R}^+$, $f^+(x + y) = f^+(x) + f^+(y)$ for all $x$ and $y$ in $\mathbb{R}^+$. Thus $f^+$ satisfies the hypothesis of Lemma 1.3. Therefore, by Lemma 1.3 let $c$ in $\mathbb{R}^+$ be such that for all $x$ in $\mathbb{R}^+$, $f(x) = f^+(x) = c \cdot x$. Because $f(0) = 0$, $f(0) = c \cdot 0$. Because for positive $u$

$$0 = f(0) = f(u + (-u)) = f(u) + f(-u),$$

$f(-u) = -f(u)$. Thus it follows that for negative $x$,

$$f(x) = -f(-x) = -c \cdot (-x) = c \cdot x. \quad □$$

Lemma 1.3 and 1.4 are versions of a famous theorem of the 19th-century mathematician Cauchy. Cauchy showed that the only solutions for a continuous function $f$ for which the equation,

$$f(x + y) = f(x) + f(y), \qquad (1.14)$$

holds for all real $x$ and $y$ was $f$ = multiplication by a positive real. Equation 1.14 is today called "Cauchy's Equation."

**Lemma 1.5** *Suppose $f$ and $g$ are strictly increasing functions from $\mathbb{R}$ onto $\mathbb{R}$ such that for all $x$ and $y$ in $\mathbb{R}$,*

$$f(x+y) = f(x) + g(y). \tag{1.15}$$

*Then for some positive real numbers $c$ and $d$, $f(x) = c \cdot x + d$ and $g(y) = cy$.*

**Proof.** Because $x + y = y + x$, it follows from Equation 1.15 that

$$f(x) + g(y) = f(y) + g(x). \tag{1.16}$$

Substituting $x + y$ for $x$ and $z$ for $y$ in Equation 1.15 yields

$$f[(x+y) + z] = f(x+y) + g(z).$$

Then by Equation 1.16,

$$f[(x+y) + z] = f(z) + g(x+y).$$

Thus by Equations 1.16 and 1.15,

$$f(z) + g(x+y) = f(x+y) + g(z) = f(x) + g(y) + g(z). \tag{1.17}$$

Equation 1.17 is valid for all $x$, $y$, and $z$. Taking $z = x$ yields,

$$f(x) + g(x+y) = f(x) + g(y) + g(x),$$

which yields,

$$g(x+y) = g(x) + g(y).$$

Thus by Lemma 1.4, let $c$ in $\mathbb{R}^+$ be such that

$$g(y) = cy.$$

Letting $x = 0$ in Equation 1.15 and $d = f(0)$ then yields,

$$f(y) = cy + d. \quad \square$$

**Lemma 1.6 (Proof of Lemma 1.1)** *Suppose $\Psi$ and $K$ are strictly increasing functions from $\mathbb{R}^+$ onto $\mathbb{R}^+$ such that for all $\beta$ and $r$ in $\mathbb{R}^+$,*

$$\Psi(\beta r) = K(\beta)\Psi(r). \tag{1.18}$$

*Then for some positive $\lambda$ and $\gamma$, $\Psi(r) = \lambda r^\gamma$.*

**Proof.** Taking logarithms of both sides of Equation 1.18 yields,

$$\log[\Psi(\beta r)] = \log[K(\beta)] + \log[\Psi(r)]. \tag{1.19}$$

Define the functions $f$ and $g$ on $\mathbb{R}$ as follows: For each $u$ in $\mathbb{R}^+$,
$$f(\log(u)) = \log[\Psi(u)] \text{ and } g(\log(u)) = \log[K(u)].$$

It easily follows that $f$ and $g$ are strictly increasing functions from $\mathbb{R}$ onto $\mathbb{R}$. Let $r$ and $\beta$ be arbitrary elements of $\mathbb{R}^+$ and
$$x = \log(r) \text{ and } y = \log(\beta).$$

It then follows from Equation 1.19 that
$$\begin{aligned} f(x+y) &= f[\log(r) + \log(\beta)] = f[\log(r\beta)] = f[\log(\beta r)] \\ &= \log[\Psi(\beta r)] = \log[K(\beta)] + \log[\Psi(r)] \\ &= g(\log(\beta)) + f(\log(r)) \\ &= g(y) + f(x) = f(x) + g(y). \end{aligned}$$

By Lemma 1.5, let $c$ and $d$ be reals such that $f(x) = cx + d$ and $g(y) = cy$. Then,
$$\begin{aligned} \Psi(r) &= e^{\log[\Psi(r)]} = e^{f(\log(r))} = e^{f(x)} = e^{cx+d} \\ &= e^{c\log(r)+d} = e^d \cdot e^{\log(r^c)} = e^d \cdot r^c = \lambda r^\gamma, \end{aligned}$$

where $\lambda = e^d$ and $\gamma = c$. □

# Part I
# Measurement

## Chapter 2

# Representational Measurement Theory

## 2.1 INTRODUCTION

*Measurement* is the means for representing empirical or qualitative entities and relationships in mathematics. Once represented, the full power of mathematics and its wealth of known results and concepts can brought to bear for understanding the represented relationships and their implications. Some of these implications have empirical or qualitative consequences for the fragment of science under consideration. These are called *meaningful.* Others are intractably meshed with mathematics and cannot be identified with empirical or qualitative objects or relations or with empirical or qualitative propositions about the fragment of science under consideration. These are called *meaningless.* And still others are about a different fragment of science. These are also called *meaningless,* although from a different viewpoint, for example, the viewpoint of the other fragment of science, they may be meaningful. Thus *meaningfulness,* as used throughout this book, is a relative concept depending on the fragment of science under consideration. In particular, "meaningless" should not be interpreted as "devoid of meaning;" instead it should be interpreted as "not part of the fragment of science under consideration."

In the literature, there are various theories as to what constitute measurement and meaningfulness. Some are more formal than others. The theory of measurement employed throughout most of this book is the *Representational Theory of Measurement,* or the *representational theory* for short. It is one of a few competing theories as to what constitutes "measurement." The theory of meaningfulness employed is a new theory founded on ideas presented in Narens (2002a, 2002b); it is also one of the few competing theories of meaningfulness.

Representation and uniqueness theorems for various measurement structures are presented throughout this chapter and book. Because these theorems are well known in measurement literature, most of their proofs are referred to appropriate places in literature.

## 2.2 PRELIMINARIES

### 2.2.1 Basic Notation

Before proceeding to the representational theory, some preliminary definitions and theorems are needed. This section provides these.

Throughout this book the following notation is employed: $\mathbb{R}$ denotes the set of reals, $\mathbb{R}^+$ the set of positive reals, $\mathbb{I}$ the set of integers, and $\mathbb{I}^+$ the set of positive integers. $*$ is the operation of function composition, that is, $f * g(x) = f[g(x)]$. $\in$ is the relation of set membership, $\subseteq$ the subset relation, $\subset$ the proper subset relation, and $\varnothing$ is the empty set. $\square$

### 2.2.2 Isomorphisms

**Definition 2.1 (relational structure)** $\mathfrak{X} = \langle X, R_j \rangle_{j \in J}$ is said to be a *relational structure* if and only if $X$ is a non-empty set, called the *domain* of $\mathfrak{X}$, and for each $j$ in $J$, $R_j$ is an element of $X$ (that is, a *0-ary relation* on $X$), or is a subset of elements of $X$ (that is, a *1-ary* relation on $X$), or is a *n-ary* relation on $X$, $n \geq 2$. $X$ and $R_j$, $j \in J$, are called the *primitives* of $\mathfrak{X}$. $\square$

In the previous definition, a function $y = f(x)$ from $X$ into $X$ is considered to be the *2-ary* (or *binary*) relation $\{(x, f(x)) \,|\, x \in X\}$, where the notation $(u, v)$ denotes the ordered pair of elements $u$ and $v$.

The reader should pay attention to the following definition and convention. They are used throughout the book.

**Definition 2.2** Let $\varphi$ be a one-to-one function from the non-empty set $X$ onto the non-empty set $Y$. Then $\varphi$ is extended to $n$-ary relations, $n \geq 1$, as follows: For each subset $A$ of $X$,

$$\varphi(A) = \{\varphi(x) \,|\, x \in A\}.$$

For each $n$-ary relation $R$ on $X$, $n > 1$, $\varphi(R) = S$, where $S$ is the $n$-ary relation on $Y$ such that for all $y_1, \ldots, y_n$ in $Y$,

$$S(y_1, \ldots, y_n) \text{ iff } R(\varphi^{-1}(y_1), \ldots, \varphi^{-1}(y_n)). \quad \square$$

**Convention 2.1** Let $\varphi$ be a one-to-one function from the non-empty set $X$ onto the non-empty set $Y$.

Let $h$ be a function from $X$ into $X$. Then $h$ is a set of ordered pairs, $\{(x, h(x)) \mid x \in X\}$ that is a binary relation on $X$. By Definition 2.2, $\varphi(h) = \{(\varphi(x), \varphi(h(x)) \mid x \in X\}$. Translated into function notation this yields the following: For all $x$ in $X$,

$$\varphi(h)(\varphi(x)) = \varphi(h(x)). \tag{2.1}$$

More generally, let $R$ be a $n$-ary, $n \geq 1$, relation on $X$. Then by definition, $\varphi(R)$ is the relation $S$ such that for all $y_1, \ldots, y_n$, $S(y_1, \ldots, y_n)$ holds if and only if for some $x_1, \ldots, x_n$, $R(x_1, \ldots, x_n)$ holds and $\varphi(x_1) = y_1, \ldots, \varphi(x_n) = y_n$. Thus for all $u_1, \ldots, u_n$ in $X$,

$$R(u_1, \ldots, u_n) \text{ iff } \varphi(R)(\varphi(u_1), \ldots, \varphi(u_n)). \quad \square$$

**Theorem 2.1** *Let $\varphi$ be a one-to-one function from the non-empty set $X$ onto the non-empty set $Y$ and $f$ and $g$ be arbitrary functions from $X$ onto $X$. Then*

$$\varphi(f * g) = \varphi(f) * \varphi(g).$$

**Proof.** By repeated application of Equation 2.1, for each $x$ in $X$,

$$\begin{aligned} \varphi(f * g)(x) &= \varphi(f[g(x)]) \\ &= \varphi(f)[\varphi(g(x))] \\ &= \varphi(f)[\varphi(g)[\varphi(x)]] \\ &= (\varphi(f) * \varphi(g))[\varphi(x)]. \quad \square \end{aligned}$$

**Definition 2.3 (isomorphism)** Let $\mathfrak{X} = \langle X, R_j \rangle_{j \in J}$ and $\mathfrak{Y} = \langle Y, S_k \rangle_{k \in K}$ be relational structures. Then $\varphi$ is said to be an *isomorphism from $\mathfrak{X}$ onto $\mathfrak{Y}$* if and only if $\varphi$ is a one-to-one function, $\varphi(X) = Y$, $J = K$, and for all $j$ in $J$, $\varphi(R_j) = S_j$.

$\mathfrak{X}$ and $\mathfrak{Y}$ are said to be *isomorphic* if and only if there exists an isomorphism from $\mathfrak{X}$ onto $\mathfrak{Y}$. $\quad \square$

The following theorem is immediate from Definition 2.3.

**Theorem 2.2** *Let $\mathfrak{X}$, $\mathfrak{Y}$, and $\mathfrak{Z}$ be relational structures, $\varphi$ be an isomorphism from $\mathfrak{X}$ onto $\mathfrak{Y}$, and $\gamma$ be an isomorphism from $\mathfrak{Y}$ onto $\mathfrak{Z}$. Then the following two statements hold:*

1. *$\varphi^{-1}$ is an isomorphism from $\mathfrak{Y}$ onto $\mathfrak{X}$.*

2. *$\gamma * \varphi$ is an isomorphism from $\mathfrak{X}$ onto $\mathfrak{Z}$.* $\quad \square$

**Definition 2.4 (symmetry)** Let $\mathfrak{X}$ be a relational structure. Then $\alpha$ is said to be a *symmetry* of $\mathfrak{X}$ if and only if $\alpha$ is an isomorphism of $\mathfrak{X}$ onto itself. $\quad \square$

The concept of "symmetry" plays a major role throughout this book. The following theorem characterizes the symmetries of the relational structure $\langle \mathbb{R}^+, \leq, + \rangle$.

**Theorem 2.3** *For each $r$ in $\mathbb{R}^+$, let $\alpha_r(x)$ be $r \cdot x$. Then the set of symmetries of $\langle \mathbb{R}^+, \leq, + \rangle = \{\alpha_r \mid r \in \mathbb{R}^+\}$.*

**Proof.** It is immediate that for each $r$ in $\mathbb{R}^+$, $\alpha_r$ is a symmetry of $\langle \mathbb{R}^+, \leq, + \rangle$. Suppose $\alpha$ is a symmetry of $\langle \mathbb{R}^+, \leq, + \rangle$. Then, because $\alpha$ is a symmetry,
$$x \leq y \text{ iff } \alpha(x) \leq \alpha(y),$$
for all $x$ and $y$ in $\mathbb{R}^+$. Thus $\alpha$ is a strictly increasing function from $\mathbb{R}^+$ onto $\mathbb{R}^+$. Because $\alpha$ is a symmetry,
$$\alpha(x + y) = \alpha(x) + \alpha(y)$$
for all $x$ and $y$ in $\mathbb{R}^+$. Therefore by Lemma 1.3, $\alpha = \alpha_r$ for some $r$ in $\mathbb{R}^+$. □

**Theorem 2.4** *Let $\mathfrak{X} = \langle X, R_j \rangle_{j \in J}$ and $\mathfrak{Y} = \langle Y, S_j \rangle_{j \in J}$ be relational structures and $\varphi$ be an isomorphism from $\mathfrak{X}$ onto $\mathfrak{Y}$. Then for each symmetry $\alpha$ of $\mathfrak{X}$, $\varphi(\alpha)$ is a symmetry of $\mathfrak{Y}$.*

**Proof.** Let $\theta = \varphi(\alpha)$, $y_1, \ldots, y_n$ be arbitrary elements of the domain of $\mathfrak{Y}$, $x_1 = \varphi^{-1}(y_1), \ldots, x_n = \varphi^{-1}(y_n)$, and $j$ be an arbitrary element of $J$. Then

$$\begin{aligned}
S_j(y_1, \ldots, y_n) \quad &\text{iff} \quad R_j(x_1, \ldots, x_n) \\
&\text{iff} \quad R_j(\alpha(x_1), \ldots, \alpha(x_n)) \\
&\text{iff} \quad S_j(\varphi(\alpha(x_1)), \ldots, \varphi(\alpha(x_n))) \\
&\text{iff} \quad S_j([\varphi(\alpha)](\varphi(x_1)), \ldots, [\varphi(\alpha)](\varphi(x_n))) \quad \text{(Equation 2.1)} \\
&\text{iff} \quad S_j(\theta(y_1), \ldots, \theta(y_n)). \quad \square
\end{aligned}$$

## 2.3 CANTOR'S CHARACTERIZATION OF THE CONTINUUM

The form of measurement employed throughout this book concerns situations where the set of qualitative objects $X$ under consideration are ordered by the binary relation $\preceq$ such that there is an isomorphism $\varphi$ from $\langle X, \preceq \rangle$ onto $\langle \mathbb{R}^+, \leq \rangle$. This is a special case of the representational theory, where $\mathfrak{X} = \langle X, \preceq \rangle$ is a qualitative structure of objects to be measured, $\mathfrak{N} = \langle \mathbb{R}^+, \leq \rangle$ is the numerical system in which objects of $X$ are measured, and $\varphi$ is a measuring function that assigns numbers to objects. In the representational theory, axioms are given about $\mathfrak{X}$ such that it can be shown

# Representational Measurement Theory

that there exists an isomorphism of $\mathfrak{X}$ onto $\mathfrak{N}$. These axioms are to be formulated in terms of $X$, $\preceq$, and concepts defined in terms of them. There is a classical axiomatization due to Cantor (1895) that accomplishes this. In order to give this axiomatization, some notation needs to be made explicit and some basic concepts need to be defined.

**Definition 2.5 (denumerable set)** A set $Y$ is said to be *denumerable* if and only if there exists a one-to-one function $f$ from $\mathbb{I}^+$ onto $Y$. □

**Definition 2.6 (total ordering)** $\preceq$ is said to be a *total ordering* on $X$ if and only if $X$ is a non-empty set and $\preceq$ is a binary relation on $X$ that is

- *transitive* (for all $x$, $y$ and $z$ in $X$, if $x \preceq y$ and $y \preceq z$, then $x \preceq z$),

- *connected* (either $x \preceq y$ or $y \preceq x$ for all $x$ and $y$ in $X$),

- and *antisymmetric* (for all $x$ and $y$ in $X$, if $x \preceq y$ and $y \preceq x$, then $x = y$).

Suppose $\preceq$ is a total ordering on $X$. Then it easily follows that $x \preceq y$ and $y \preceq x$ implies $x = y$. By definition, $x \prec y$ if and only if $x \preceq y$ and $x \neq y$.

By definition, $\langle X, \preceq \rangle$ is said to be a *totally ordered set* if and only if $\preceq$ is a total ordering on $X$. □

**Definition 2.7** Let $\langle X, \preceq \rangle$ be a totally ordered set and $Y$ a non-empty subset of $X$. Then a function $f$ from $Y$ into $X$ is said to be $\preceq$-*strictly increasing* if and only if for all $x$ and $y$ in $Y$, if $x \prec y$ then $f(x) \prec f(y)$. And $Y$ is said to be $\preceq$-*bounded above in* $X$ if and only if there exists $z$ in $X$ such that for all $y$ in $Y$, $y \preceq z$. □

**Definition 2.8 (Dedekind completeness)** Let $\mathfrak{X} = \langle X, \preceq \rangle$ be a totally ordered set. Then $\mathfrak{X}$ is said to be *Dedekind complete* if and only if for each non-empty subset $Z$ of $X$ that is $\preceq$-bounded above in $X$, there exists an element $a$ in $X$, called the *supremuim of $Z$ (in $X$)*, or sup $Z$ for short, such that

(i) $z \preceq a$ for all $z$ in $Z$, and

(ii) if $b$ in $X$ is such that $z \preceq b$ for all $z$ in $Z$, then $a \preceq b$. □

**Definition 2.9 (continuum)** $\langle X, \preceq \rangle$ is said to be a *continuum* if and only if the following four statements hold:

1. *Total ordering*: $\preceq$ is a total ordering on $X$ (Definition 2.6).

2. *Unboundedness*. $\langle X, \preceq \rangle$ has no $\preceq$-greatest or $\preceq$-least element.

3. *Denumerable density*: There exists a denumerable subset $Y$ of $X$ such that for each $x$ and $z$ in $X$, if $x \prec z$ then there exists $y$ in $Y$ such that $x \prec y$ and $y \prec z$.

4. *Dedekind completeness*: $\langle X, \preceq \rangle$ Dedekind complete (Definition 2.8). □

**Theorem 2.5 (Cantor's Continuum Theorem)** $\mathfrak{X} = \langle X, \preceq \rangle$ is a continuum if and only if $\mathfrak{X}$ is isomorphic to $\langle \mathbb{R}^+, \leq \rangle$.

**Proof.** Cantor (1895). (A proof is also given in Theorem 2.2.2 of Narens, 1985.)

The following is an immediate consequence of Theorem 2.5.

**Theorem 2.6** Let $\mathfrak{X} = \langle X, \preceq \rangle$ be a continuum and $\mathcal{F}$ be the set of isomorphisms of $\mathfrak{X}$ onto $\langle \mathbb{R}^+, \leq \rangle$. Then for each $f$ in $\mathcal{F}$,

$$\mathcal{F} = \{g * f \mid g \text{ is a strictly increasing function from } \mathbb{R}^+ \text{ onto } \mathbb{R}^+\}. \quad \square$$

In the representational theory, Theorem 2.5 is called an *existence theorem*, because it shows the existence of an isomorphism, and Theorem 2.6 is called a *uniqueness theorem*, because it characterizes how any two isomorphisms are related.

Note that it follows from the definition of continuum (Definition 2.9) that $\langle \mathbb{R}^+, \leq \rangle$ is a continuum. In general $\langle N, \leq \rangle$ is a continuum, where $N$ is a non-empty interval of reals without endpoints, including cases where $N$ is an infinite interval without endpoints. In particular, $\langle \mathbb{R}, \leq \rangle$ is a continuum. Thus it follows from Theorems 2.6 and 2.2 that each continuum $\langle X, \preceq \rangle$ is isomorphic to $\langle \mathbb{R}, \leq \rangle$.

## 2.4 CONTINUOUS STRUCTURES

In the scientific literature, the term "scale" denotes a specific way of measuring, as in "the centimeter scale," as well as a set of specific ways of measuring, as in "a ratio scale for length." Throughout this book, the terms "scale" and "scale family" refer to a set of specific ways of measuring.

**Definition 2.10 (scale family)** Let $Y$ be a non-empty set. A *scale family* (or *scale*) on $Y$ is a non-empty set of functions from $Y$ onto a subset of $\mathbb{R}$. Let $\mathcal{F}$ be a scale family on $Y$. Then the elements of $\mathcal{F}$ are called *measuring functions*. □

The following three kinds of scales are routinely applied in the behavioral sciences.

**Definition 2.11 (ratio, interval, and ordinal scales)** $\mathcal{F}$ is said to be a *ratio scale* if and only if (1) $\mathcal{F}$ is a scale family of functions onto $\mathbb{R}^+$, and (2) for any $f$ in $\mathcal{F}$,
$$\mathcal{F} = \{rf \,|\, r \in \mathbb{R}^+\}.$$

$\mathcal{F}$ is said to be an *interval scale* if and only if (1) $\mathcal{F}$ is a scale family of functions onto $\mathbb{R}$, and (2) for any $f$ in $\mathcal{F}$,
$$\mathcal{F} = \{rf + s \,|\, r \in \mathbb{R}^+ \text{ and } s \in \mathbb{R}\}.$$

$\mathcal{F}$ is said to be an *ordinal scale* if and only if (1) $\mathcal{F}$ is a scale family of functions onto $\mathbb{R}^+$, and (2) for any $f$ in $\mathcal{F}$,
$$\mathcal{F} = \{g * f \,|\, g \text{ is a strictly increasing function from } \mathbb{R}^+ \text{ onto } \mathbb{R}^+\}. \quad \square$$

For the purposes of this book, the concepts of "ratio scale" and "ordinal scale" are limited to situations where the scale families are onto $\mathbb{R}^+$. The literature often defines them to include an additional case where the scale family is onto $\mathbb{R}$. The ideas developed in this book easily extend to this additional case. However, for extending "ratio scale" to $\mathbb{R}$, concepts sometimes need to be amended to accommodate the special role 0 plays in such a scale family.

The theory of measurement for continua is a special case of the representational theory. A more general case, and the one that is the focus of this book, is the theory of measurement for continuous structures.

**Definition 2.12 (continuous structure)** $\mathfrak{X} = \langle X, \preceq, R_j, \rangle_{j \in J}$ is said to be a *continuous structure* if and only if $\langle X, \preceq \rangle$ is a continuum (Definition 2.9) and $\mathfrak{X}$ is a relational structure (Definition 2.1). $\quad \square$

**Theorem 2.7** *Suppose $\mathfrak{X} = \langle X, \preceq, R_j, \rangle_{j \in J}$ is a continuous structure. Then there exist a numerical structure $\mathfrak{N} = \langle \mathbb{R}^+, \leq, R_j^\star \rangle_{j \in J}$ and an isomorphism of $\mathfrak{X}$ onto $\mathfrak{N}$.*

**Proof.** By Theorem 2.5, let $\varphi$ be an isomorphism of $\langle X, \preceq \rangle$ onto $\langle \mathbb{R}^+, \leq \rangle$. For each $j$ in $J$, let
$$R_j^\star = \varphi(R_j),$$
where $\varphi(R_j)$ is as defined in Definition 2.2. Then by Definition 2.3, $\varphi$ is an isomorphism of $\mathfrak{X}$ onto $\mathfrak{N} = \langle \mathbb{R}^+, \leq, R_j^\star \rangle$. $\quad \square$

**Definition 2.13** A *theory of measurement* consists of a precise specification of how a scale family $\mathcal{F}$ of functions is formed. The currently dominant approach to measurement in the literature is the *representational theory*. For the purposes of this book, the representational theory is formulated as follows: Scale families $\mathcal{F}$ on a set $Y$ result by providing a relational structure $\mathfrak{Y}$ with domain $Y$ and a numerical structure $\mathfrak{N}$ with domain a subset of $\mathbb{R}$ such that $\mathcal{F}$ is the set of isomorphisms of $\mathfrak{Y}$ onto $\mathfrak{N}$.

Let $\mathcal{F}$ be a scale family that results by the representational theory. In the representational theory, elements of $\mathcal{F}$ are usually called *representations*. Throughout this book elements of $\mathcal{F}$ are instead usually called "measuring functions" or "isomorphisms." □

In his derivation of the psychophysical power law, Plateau used a "midway" operation (Section 1.1). The following definition provides an algebraic characterization of "midway."

**Definition 2.14 (continuous bisection structure)** $\langle X, \preceq, \ominus \rangle$ is said to be a *continuous bisection structure with bisection operation* $\ominus$ if and only if $\ominus$ is a binary operation on $X$ and the following six conditions hold for all $u$, $x$, $y$, and $z$ in $X$:

1. *Continuum:* $\langle X, \preceq \rangle$ is a continuum.

2. *Idempotence:* $u \ominus u = u$.

3. *Commutativity:* $x \ominus y = y \ominus x$.

4. *Solvability:* There exists $v$ in $X$ such that $u \ominus x = v \ominus y$.

5. *Monotonicity:* $x \preceq y$ if and only if $x \ominus z \preceq y \ominus z$.

6. *Bisymmetry:* $(u \ominus x) \ominus (y \ominus z) = (u \ominus y) \ominus (x \ominus z)$. □

**Theorem 2.8** *Suppose* $\mathfrak{X} = \langle X, \preceq, \ominus \rangle$ *is a continuous bisection structure. Then (1)* $\mathfrak{X}$ *and* $\mathfrak{N} = \langle \mathbb{R}^+, \leq, \ominus^\star \rangle$ *are isomorphic, where* $\ominus^\star$ *is the binary operation on* $\mathbb{R}$ *such that for all* $r$ *and* $s$ *in* $\mathbb{R}$,

$$r \ominus^\star s = \frac{r+s}{2},$$

*and (2) the set of isomorphisms of* $\mathfrak{X}$ *onto* $\mathfrak{N}$ *form an interval scale.*

**Proof.** Theorem 10 of Section 6.9 of Krantz, Luce, Suppes, and Tversky (1971) and Theorem 2.1 of Luce and Narens (1985).[2] □

---

[2]The first four axioms of Theorem 10 of Section 6.9 of Krantz et al. (1971) easily follow from the Conditions 1 to 6 of Definition 2.14. The remaining axiom (the Archimedean axiom) follows from Theorem 2.1 of Luce and Narens (1985).

Theorem 10 of Section 6.9 of Krantz et al. (1971) produces a measuring function $\varphi$ that is into $\mathbb{R}$. However, Solvability is a stronger condition than the related axiom Krantz, et al. (1971), and it is easy to apply the conditions of Solvability and Continuum to $\varphi$ to show that $\varphi$ is onto $\mathbb{R}$.

Consider a case similar to Plateau's experiment (Section 1.1) where an observer is presented gray disks $x$ and $y$ and is asked to produce a disk gray disk $z$ that is midway between them in terms of grayness of $x$ and $y$. This produces an operation $\ominus$ on gray disks defined by $x \ominus y = z$. Let $x \preceq y$ stand for the disk $x$ is a lighter gray than $y$ or $x$ and $y$ have the same amount of gray-lightness. Then intuitively the conditions for a continuous bisection structure should hold for gray disks, except possibly for bisymmetry and solvability. Bisymmetry,

$$(u \ominus x) \ominus (y \ominus z) = (u \ominus y) \ominus (x \ominus z), \qquad (2.2)$$

corresponds to an experiment that predicts the disk produced by the left side of Equation 2.2 is identical in the appearance of grayness to the disk produced by the right side. It is one way of capturing the qualitative essence of the concept of "midway." Thus bisymmetry is a key condition to test experimentally to see if the observer is truly producing midway grays. Plateau's lack of a test of bisymmetry, or a similar kind of condition that captures the mathematical structure of "midway," shows his argument for the power law to be conceptually and methodologically unsound. The impact of solvability is discussed in Section 6.1.

Helmholtz (1887) and Hölder (1901) provided qualitative theories for fundamental physical qualities like length, mass, time, et cetera. (Hölder's theory is a more general and mathematically rigorous version of Helmholtz's.) Their theories are based on the concept of a "continuous extensive structure."

**Definition 2.15 (continuous extensive structure)** $\mathfrak{X} = \langle X, \preceq, \oplus \rangle$ is said to be a *continuous extensive structure* if and only if the following seven axioms hold:

1. *Total Ordering*: $\preceq$ is a total ordering on $X$ (Definition 2.6).

2. *Density*: For all $x$ and $z$ in $X$, if $x \prec z$ then for some $y$ in $X$, $x \prec y \prec z$.

3. *Associativity*: $\oplus$ is a binary operation that is *associative*; that is,

$$(x \oplus y) \oplus z = x \oplus (y \oplus z)$$

   for all $x$, $y$, and $z$ in $X$.

4. *Monotonicity*: For all $x$, $y$, and $z$ in $X$,

$$x \preceq y \text{ iff } x \oplus z \preceq y \oplus z \text{ iff } z \oplus x \preceq z \oplus y.$$

5. *Solvability*: For all $x$ and $y$ in $X$, if $x \prec y$, then for some $z$ in $X$, $x \oplus z = y$.

6. *Positivity*: $x \prec x \oplus y$ and $y \prec x \oplus y$, for all $x$ and $y$ in $X$.

7. *Dedekind Completeness*: $\langle X, \preceq \rangle$ is Dedekind complete (Definition 2.8). □

In essence Helmholtz (1887) showed the following theorem.

**Theorem 2.9** *Suppose* $\mathfrak{X} = \langle X, \preceq, \oplus \rangle$ *is a continuous extensive structure. Then the set* $S$ *of isomorphisms of* $\mathfrak{X}$ *onto* $\langle \mathbb{R}^+, \leq, + \rangle$ *is a ratio scale.*

Proofs of generalizations of Theorem 2.9 can be found in Chapter 3 of Krantz et al. (1971) and in Section 9 of Chapter 2 of Narens (1985).[3] □

The measurement of mass by use of a equal arm pan balance is an example of continuous extensive measurement. Physical objects $a$ and $b$ are said to be "equivalent in mass," $a \sim b$, if and only if when placed in opposite pans $a$ balances $b$. It is assumed that $\sim$ is an equivalence relation. Let $X$ be the set of $\sim$–equivalence classes of physical objects. The binary relation $\preceq$ can be defined on $X$ as follows: $\beta \preceq \alpha$ if and only if there exist physical objects $x$ in $\alpha$ and $y$ in $\beta$ such that if $x$ and $y$ are placed in opposite pans, either they balance or the pan with $x$ becomes lower than the one with $y$. $\oplus$ is defined on $X$ as follows: $\alpha \oplus \beta = \gamma$ if and only if there exist $x$ in $\alpha$, $y$ in $\beta$, and $z$ in $\gamma$ such that when $x$ and $y$ are placed in the same pan and $z$ in the opposite, the result balances. It is assumed that $\langle X, \preceq, \oplus \rangle$ satisfies the axioms of a continuous extensive structure.

Another example of continuous extensive measurement is the measurement of length. Here $R$ is the set of measuring rods, which ideally look like line segments. Rods $a$ and $b$ are said to be equivalent in length, $a \sim b$, if and only if $a$ and $b$ can be laid side by side with endpoints exactly corresponding. It is assumed that $\sim$ is an equivalence relation. Let $X$ be the set of $\sim$–equivalence classes of elements of $R$. Then $\preceq$ can be defined on $X$ as follows: $\beta \preceq \alpha$ if and only if there exist $x$ in $\alpha$ and $y$ in $\beta$ such that either $x \sim y$ or when $x$ and $y$ are placed side by side with left endpoints exactly corresponding, then the right endpoint of $x$ extends beyond the right endpoint of $y$. $\oplus$ is defined on $X$ as follows: $\alpha \oplus \beta = \gamma$ if and only if there exist $x$ in $\alpha$, $y$ in $\beta$, and $z$ in $\gamma$ such that when $x$ and $y$ are placed on an oriented line with the right endpoint of $x$ touching the left endpoint of $y$ (i.e., "x is

---

[3]To obtain a proof of Theorem 2.9 from these generalizations, use Solvability and Dedekind Completeness to show that $X$ satisfies the "Archimedean axiom," and use simple consequences of the axioms to show that the relevant measuring functions are onto $\mathbb{R}^+$.

abutted to y"), they form a rod $w$ such that $w \sim z$. In theoretical classical physics, it is assumed that $\langle X, \preceq, \oplus \rangle$ satisfies the axioms of a continuous extensive structure.

Of course, in modern physics mass and distance are no longer measured using pan balances and measuring rods. Instead, rather sophisticated instrumentations are employed for their measurements. Nevertheless, the justifications for the correctness of the resulting measurements rest heavily on physical theory, which in turn, assumes a theory of measurement for mass and length. Measurement using continuous extensive structures is useful as a theory of measurement for *theoretical* physics, because it justifies how numbers are assigned to ideal physical entities. It is important in physical theory to make explicit its measurement assumptions, for, as is shown repeatedly throughout this book, the rules by which numbers are assigned to physical entities necessarily condition the mathematical form of physical laws.

**Convention 2.2** Throughout this book it is assumed that physical variables of interest are measured through ratio scales based on qualitative continuous extensive structures. □

This convention is justified because of results of measurement theory showing physical dimensions are divided into two types: ($i$) fundamental physical dimensions, which are measurable through ratio scales defined on continuous extensive structures on the dimensions, and ($ii$) derived physical dimensions, which are cartesian products of fundamental physical dimensions, and are measured by derived ratio scales whose measuring functions are products of ratios and products of measuring functions of the fundamental physical dimensions. For example, a measuring function for the derived dimension of acceleration is the ratio of a measuring function for the fundamental dimension of distance with the product of a measuring function for the fundamental dimension of time with itself. It is known result of measurement theory that for each derived physical dimension $X$, an ordering $\preceq$ and an extensive operation $\oplus$ can be explicitly defined in terms of the orderings and extensive operations on the fundamental physical dimensions so that the derived ratio scale used to measure $X$ coincides with the set of isomorphisms of $\langle X, \preceq, \oplus \rangle$ onto $\langle \mathbb{R}^+, \leq, + \rangle$.

Many kinds of continuous structures have been axiomatized in the literature. The compactness of this book precludes descriptions of the vast majority of these. Krantz, et al. (1971) provides a number of examples and scientific applications of continuous and other kinds of measurement structures. At a more advanced mathematical level, Luce and Narens (1985) describe a wide class of measurement structures that generalize in various ways extensive and bisection structures.

## 2.5 SCALE TYPES

The following two scale types are used to measure various important situations.

**Definition 2.16 (log-Interval scale)** $\mathcal{F}$ is said to be a *log-interval scale* if and only if (1) $\mathcal{F}$ is a scale family of functions that are onto $\mathbb{R}^+$, and (2) for any $f$ in $\mathcal{F}$,
$$\mathcal{F} = \{rf^s \mid r \in \mathbb{R}^+ \text{ and } s \in \mathbb{R}^+\}. \quad \square$$

**Definition 2.17 (translation scale)** $\mathcal{F}$ is said to be a *translation scale* if and only if (1) $\mathcal{F}$ is a scale family of functions that are onto $\mathbb{R}$, and (2) for any $f$ in $\mathcal{F}$,
$$\mathcal{F} = \{f + r \mid r \in \mathbb{R}\}. \quad \square$$

Interval scales and log-interval scales are variants of one another in the following sense:

Suppose $\mathfrak{X} = \langle X, \preceq, R_j, \rangle_{j \in J}$ is a qualitative structure,
$$\mathfrak{N} = \langle \mathbb{R}, \leq, R_j^\star \rangle_{j \in J}$$
is a numerical representing structure, and $\mathcal{S}$ is an interval scale of isomorphisms from $\mathfrak{X}$ onto $\mathfrak{N}$. Let $\varphi$ be an arbitrary element of $\mathcal{S}$. For each $z$ in $X$, let
$$\psi(z) = e^{\varphi(z)},$$
and for each $j$ in $J$, if $R_j$ is a $n$-ary relation on $X$, let $R_j'$ be the $n$-relation on $\mathbb{R}^+$ such that for all $z_1, \ldots, z_n$ in $X$,
$$R_j^\star(\varphi(z_1), \ldots, \varphi(z_n)) \text{ iff } R_j'(e^{\varphi(z_1)}, \ldots, e^{\varphi(z_n)}).$$
Let $\mathfrak{N}' = \langle \mathbb{R}^+, \leq, R_j' \rangle_{j \in J}$. Then,
$$\begin{aligned} R_j(z_1, \ldots, z_n) \quad &\text{iff} \quad R_j^\star(\varphi(z_1), \ldots, \varphi(z_n)) \\ &\text{iff} \quad R_j'(e^{\varphi(z_1)}, \ldots, e^{\varphi(z_n)}) \\ &\text{iff} \quad R_j'(\psi(z_1), \ldots, \psi(z_n)). \end{aligned}$$
Also, for all $x$ and $y$ in $X$,
$$x \preceq y \text{ iff } \varphi(x) \leq \varphi(y) \text{ iff } e^{\varphi(x)} \leq e^{\varphi(y)} \text{ iff } \psi(x) \leq \psi(y).$$

Thus $\psi$ is an isomorphism of $\mathfrak{X}$ onto $\mathfrak{N}'$. A transformation of $\varphi$ of the form, $r\varphi + s$, $r \in \mathbb{R}^+$ and $s \in \mathbb{R}$, yields a transformation $\psi'$ of $\psi$ such that for all $t$ in $\mathbb{R}^+$,
$$\psi'(t) = e^{r\varphi(t)+s} = e^s e^{r\varphi(t)} = e^s \cdot [e^{\varphi(t)}]^r = e^s \cdot \psi(t)^r.$$

It easily follows that $\psi'$ is also an isomorphism of $\mathfrak{X}$ onto $\mathfrak{N}'$ and that each isomorphism of $\mathfrak{X}$ onto $\mathfrak{N}'$ has this form for some $r$ in $\mathbb{R}^+$ and $s$ in $\mathbb{R}$. Thus the function

$$t \to e^t$$

from $\mathbb{R}$ onto $\mathbb{R}^+$ maps the interval scale $\mathcal{S}$ onto the log-interval scale $\mathcal{S}' = \{u\psi^v \mid u \in \mathbb{R}^+ \text{ and } v \in \mathbb{R}^+\}$, where $\psi$ is as previously defined.

Similarly, the function $t \to \log t$ will map a ratio scale of isomorphisms onto a translation scale of isomorphisms.

**Definition 2.18 (representational equivalence)** Scale families of measuring functions $\mathcal{S}$ and $\mathcal{S}'$ are said to be *representationally equivalent* if and only if there exists a qualitative structure $\mathfrak{X}$ such that $\mathcal{S}$ and $\mathcal{S}'$ are scale families of isomorphisms of $\mathfrak{X}$ onto some numerical structures. □

It easily follows that representational equivalence is an equivalence relation. The previous discussion and definition show that an interval scale of a continuous structure is representationally equivalent to a log-interval scale on that structure, and that a ratio scale of a continuous structure is representationally equivalent to a translation scale.

It is a principle of the representational theory of measurement that for each qualitative structure $\mathfrak{X} = \langle X, \preceq, R_j, \rangle_{j \in J}$, it is proper to use any isomorphic numerical structure $\mathfrak{N}$ to measure $\mathfrak{X}$. Other theories of measurement view things differently. For example, some measurement theorists view measurement as an empirical process of assigning numbers to objects. Empiricalness greatly restricts the measuring functions, even under idealization. In such cases, a measuring function $\varphi$ can be used to construct an numerical representing structure $\mathfrak{N} = \langle N, \leq^\star, R_j^\star, \rangle_{j \in J}$ by $\varphi(X) = N$, $\varphi(\leq) = \leq^\star$, and for each $j$ in $J$, $\varphi(R_j) = R_j^\star$, thus making $\varphi$ an isomorphism from $\mathfrak{X}$ onto $\mathfrak{N}$. However, because the representational theory does not put constructivity restrictions on the measuring functions, it can be shown that there will be numerical representing structures $\mathfrak{N}'$—even for simple cases of $\mathfrak{X}$—such that no isomorphism from $\mathfrak{X}$ onto $\mathfrak{N}'$ is constructive, that is, such that there is no practical way of producing isomorphisms from $\mathfrak{X}$ onto $\mathfrak{N}'$. This perceived fault of the representational theory can be corrected by requiring a practical way of producing the representing isomorphisms and providing a rigorous and useful definition of "practical." Although such a modification of the representational theory would capture better the real-world empirical assignments of numbers to objects, it would make other applications of measurement theory in mathematical science more cumbersome.

**Definition 2.19 (homogeneous scale)** A scale family $\mathcal{S}$ of isomorphisms from a continuous structure $\mathfrak{X}$ onto $\mathfrak{N}$ is said to be *homogeneous* if and only if for each $x$ in the domain of $\mathfrak{X}$ and each $r$ in the domain of $\mathfrak{N}$, there exists $\varphi$ in $\mathcal{S}$ such $\varphi(x) = r$. □

Ratio, interval, and ordinal scales of isomorphisms on continuous structures are homogeneous. It is natural to ask if there are other homogeneous scale families of isomorphisms from continuous structures onto numerical structures that are not representationally equivalent to a ratio, interval, or ordinal scale. The answer is that there are, but only a few that look like they are of importance to science. To understand this more fully, we employ a classification of scale types due to Narens (1981b).

**Definition 2.20** (*m*-**point homogeneity** & *n*-**point uniqueness**) Let $\mathfrak{X} = \langle X, \preceq, R_j \rangle_{j \in J}$ be a continuous structure and $\mathcal{S}$ be a scale family of isomorphisms from $\mathfrak{X}$ onto $\mathfrak{N} = \langle N, \leq, R_j^\star \rangle_{j \in J}$, $N = \mathbb{R}^+$ or $N = \mathbb{R}$. Then $\mathcal{S}$ is said to be:

1. *1-point homogeneous* if and only if it is homogeneous (Definition 2.19).

2. *2-point homogeneous* if and only if for all $x$ and $y$ in $X$ and all $u$ and $v$ in $N$, if $x \prec y$ and $u \prec v$, then for some $\varphi$ in $\mathcal{S}$, $\varphi(x) = u$ and $\varphi(y) = v$.

3. *m-point homogeneous*, $m \in \mathbb{I}^+$ and $m \geq 2$, if and only if for all $x_1, \ldots, x_m$ in $X$ and all $u_1, \ldots, u_m$ in $N$, if

   $$x_1 \prec x_2 \prec \cdots \prec x_{m-1} \prec x_m \quad \text{and} \quad u_1 \prec u_2 \prec \cdots \prec u_{m-1} \prec u_m,$$

   then there exists $\varphi$ in $\mathcal{S}$ such that

   $$\varphi(x_1) = u_1, \ \ldots, \ \varphi(x_m) = u_m.$$

4. ∞-*point homogeneous* if and only if $\mathcal{S}$ is $m$-point homogeneous for each $m \in \mathbb{I}^+$.

5. *n-point unique*, $n \in \mathbb{I}^+$, if and only if for all $\varphi$ and $\psi$ in $\mathcal{S}$, if for some $n$ distinct elements $x_1, \ldots, x_n$ of $X$,

   $$\varphi(x_1) = \psi(x_1), \ \ldots, \ \varphi(x_n) = \psi(x_n),$$

   then $\varphi = \psi$.

6. ∞-*point unique*, if and only if for each $n$ in $\mathbb{I}^+$, $\mathcal{S}$ is not $n$-point unique. □

The following is an easy consequence of Definition 2.20.

*Suppose $S$ is a scale of isomorphisms of from a continuous structure $\mathfrak{X}$ onto a numerical structure $\mathfrak{N}$ and $S$ is m-point homogeneous, $m \in \mathbb{I}^+$, and $\lambda$-point unique, $\lambda \in \mathbb{I}^+$ or $\lambda = \infty$. Then*

- $m \leq \lambda$,

- $S$ *is k-point homogeneous for each* $k \leq m$, $k \in \mathbb{I}^+$,

- *and if* $\lambda \in \mathbb{I}^+$, *then* $S$ *is p-point unique for each* $p \in \mathbb{I}^+$ *such that* $\lambda \leq p$.

**Definition 2.21 (subscale of an interval scale)** $\mathcal{T}$ is said to be a *subscale of an interval scale* if and only if $\mathcal{T}$ is a scale family such that $\mathcal{T} \subseteq \mathcal{S}$, where $\mathcal{S}$ is an interval scale. □

It easily follows that interval scales, ratio scales, and translation scales are subscales of an interval scale.

For continuous structures, ratio scales of isomorphisms are 1-point homogeneous and 1-point unique; interval scales of isomorphisms are 2-point homogeneous and 2-point unique; and ordinal scales of isomorphisms are $\infty$-point homogeneous and $\infty$-point unique. One can provide examples of scale families of continuous structures that are 1-point homogeneous and 2-point unique but not 2-point homogeneous. It can be shown that such a scale family is representationally equivalent to a subscale of an interval scale that is neither representationally equivalent a ratio scale nor to an interval scale. However, to my knowledge this kind of scale family has not played any important role in science.

Let $\mathfrak{X} = \langle X, \preceq, R_j, \rangle_{j \in J}$ be a continuous structure and $S$ be a scale of isomorphisms from $\mathfrak{X}$ onto an numerical structure $\mathfrak{N}$. Narens (1981a,b) showed that if $S$ is 1-point homogeneous and 1-point unique, then it is representationally equivalent to a ratio scale of isomorphisms; and if $S$ is 2-point homogeneous and 2-point unique, then it is representationally equivalent to an interval scale of isomorphisms. He further showed that it is impossible for $\mathfrak{X}$ to have a $m$-point homogeneous and $m$-point unique scale of isomorphisms for any positive integer $m > 2$. Alper (1985, 1987) showed that if $S$ is 1-point homogeneous and 2-point unique, then it is representationally equivalent to a subscale of an interval scale. He also showed that it is impossible for $S$ to be $m$-point homogeneous and $n$-point unique for positive integers $m$ and $n$ such that $m < n$ and $n \neq 2$. The considerations just discussed are summarized simply by the following theorem.

**Theorem 2.10 (possible homogeneous scale types)** *Let*

$$\mathfrak{X} = \langle X, \preceq, R_j \rangle_{j \in J}$$

*be a continuous structure and $\mathcal{S}$ be a homogeneous scale family of isomorphisms from $\mathfrak{X}$ onto an numerical structure $\mathfrak{N}$. Then $\mathcal{S}$ is either $\infty$-point unique or is representationally equivalent to (i) a ratio scale, (ii) an interval scale, or (iii) a subscale of a interval scale that is neither representationally equivalent to a ratio scale nor to an interval scale.* □

Stevens (1946) in a highly influential article that appeared in *Science* provided a list of scale types: "The type of scale achieved depends upon the character of the basic empirical operations performed. These operations are limited ordinarily by the nature of the thing being scaled and by our choice of procedures, but, once selected, the operations determine that there will eventuate one or another of the scales listed in Table ..." (pp. 667–668). The scale types mentioned were nominal,[4] ordinal, interval, and ratio.[5] It is interesting that he apparently believed (or guessed) that he had all the interesting homogeneous scale types. Theorem 2.10 shows that he was essentially correct: He missed the representational equivalents of homogeneous types that are subscales of an interval scale that are neither ratio scales nor interval scales—a type that appears to be of limited value to science—and some types that are homogeneous and $\infty$-point unique. An important example of the latter is the scale type that results from Fechner's method of measurement applied to a continuum of stimuli. Rather surprisingly, Stevens and later measurement theorists did not investigate this form of measurement in terms of scale type. This form of measurement is described in detail in chapter 4.

## 2.6 REPRESENTATIONAL MEANINGFULNESS

A number of different meaningfulness concepts have been associated with the representational theory, and the most of these are described in chapter 7. The following one is used for the issues covered in Part I of this book.

**Definition 2.22 (representational meaningfulness)** *Let*

---

[4] A *nominal scale* on a non-empty set is a set of one-to-one functions from the set onto a subset of $\mathbb{R}^+$. It is easy to show that a totally ordered set with a domain having 2 or more elements, for example, a continuum, cannot be representationally measured by a nominal scale of isomorphisms. For this reason, nominal scales are not developed in this book even though they are homogeneous scales.

[5] Later, Stevens added two additional scale types to his list: log-interval and *absolute*. (An absolute scale consists of a single measuring function and thus is not homogeneous if the function's domain has two or more elements.)

# Representational Measurement Theory

- $\mathfrak{X} = \langle X, \preceq, R_j \rangle_{j \in J}$ be a qualitative structure,
- $R$ be a $n$-ary relation on $X$,
- $N = \mathbb{R}^+$ or $N = \mathbb{R}$,
- $\mathfrak{N} = \langle N, \leq, R_j^\star \rangle_{j \in J}$ be a numerical representing structure,
- $\mathcal{S}$ the scale family of isomorphisms from $X$ onto $\mathfrak{N}$,
- and $\varphi \in \mathcal{S}$.

Then $R$ is said to be $\mathcal{S}$-*representationally meaningful* if and only if $\mathcal{S}$ is also the scale family of isomorphisms from

$$\langle X, \preceq, R_j, R \rangle_{j \in J} \text{ onto } \langle \mathbb{R}^+, \leq, R_j^\star, \varphi(R) \rangle_{j \in J}.$$

$R$ is said to be $\mathcal{S}$-*representationally meaningless* if and only if it is not $\mathcal{S}$-representationally meaningful. □

Because $\mathcal{S}$-representational meaningfulness requires $\mathcal{S}$ to be a scale family of isomorphisms of a qualitative structure $\mathfrak{X}$ onto a numerical representing structure $\mathfrak{N}$, it follows from Theorem 3.5 of the next chapter that a relation $R$ is representationally $\mathcal{S}$-meaningful if and only if it is representationally $\mathcal{T}$-meaningful for all scale families $\mathcal{T}$ of isomorphisms of $\mathfrak{X}$ onto some numerical representing structure $\mathfrak{M}$.

Let $\mathfrak{X} = \langle X, \preceq, R_j \rangle_{j \in J}$ be a qualitative structure, $R$ a $n$-ary relation on $X$, $N = \mathbb{R}^+$ or $N = \mathbb{R}$, $\mathfrak{N} = \langle N, \leq, R_j^\star \rangle_{j \in J}$ a numerical representing structure, and $\mathcal{S}$ the scale family of isomorphisms of $\mathfrak{X}$ onto $\mathfrak{N}$. The intuitive idea behind some measurement-theoretic meaningfulness concepts is that $R$ should somehow be "derivable" or "defined" from the primitives of $\mathfrak{X}$. The usual definability concepts of logic will not do, because science uses mathematics in its derivations in the ways that logic, as normally formulated, does not.

Suppose $R$ is added to the primitives of $\mathfrak{X}$ above and the resulting structure,

$$\langle X, \preceq, R_j, R \rangle_{j \in J},$$

is not measurable by the scale family $\mathcal{S}$. Then intuitively $R$ adds some new "information" to the primitives of $\mathfrak{X}$, because it eliminates some measuring functions from $\mathcal{S}$. Therefore $R$ is not "scientifically derivable" from the primitives of $\mathfrak{X}$, because scientific derivations, which can through measurement use all of mathematics, should not "add information" that is not already contained in the primitives of $\mathfrak{X}$. For example, if the primitives of $\mathfrak{X}$ are based on observations, then one would not want a scientific derivation to add information that is not already contained in those observations. In Part II of this book it is argued that if $R$ is scientifically derived from $\mathfrak{X}$, then it is $\mathcal{S}$-representationally meaningful. The contrapositive of this, "if $R$ is $\mathcal{S}$-representationally meaningless then it is not scientifically derivable from the primitives of $\mathfrak{X}$," has useful and important scientific applications. Thus $\mathcal{S}$-representational *meaninglessness* has some good intuitive grounding as a characterization of an important class of relations that add new information to the primitives—and various results presented throughout the book show that it is a useful concept in the conduction of science by providing criteria for excluding such relations from consideration.

$\mathcal{S}$-representational *meaningfulness* is harder to justify as a useful, intuitive concept for science. As discussed in Part II of this book, from some perspectives it is too broad of a notion for science, because it allows for the derivation of relations through highly nonconstructive means. It is formally equivalent to the foundational concept of a much honored mathematical theory called the "Erlanger Program." The Erlanger Program is described in the next chapter and is analyzed more deeply in Part II.

# Chapter 3

# Symmetries and the Erlanger Program

## 3.1 SYMMETRIES

"Symmetry" (Definition 2.4) is an important mathematical concept and a powerful multifaceted scientific tool. Nineteenth-century mathematicians and physicists discovered that the understanding of a mathematical or scientific situation in terms of its symmetries often provided profound insight into the situation. This led to the development of the modern powerful mathematical methods of inference and characterization based on symmetry. This chapter briefly describes a prominent program for the unification of qualitative and quantitative approaches to geometry that was founded on symmetries, and relates its core ideas to the representational theory of measurement.

**Convention 3.1** Throughout this section, the following notation and conventions are observed:

- $\mathfrak{X} = \langle X, \preceq, R_j \rangle_{j \in J}$ is qualitative structure.
- $G$ is the set of symmetries of $\mathfrak{X}$.
- $\mathfrak{N} = \langle N, \leq, S_j \rangle_{j \in J}$ is a representing numerical structure.
- $\mathcal{S}$ is a scale family of isomorphisms from $\mathfrak{X}$ onto $\mathfrak{N}$.
- $\varphi \in \mathcal{S}$.  $\square$

It is immediate that for each symmetry $\alpha$ of $\mathfrak{X}$, $\varphi * \alpha$ is an isomorphism of $\mathfrak{X}$ onto $\mathfrak{N}$ (because a composition of isomorphisms is again an isomorphism), and therefore $\varphi * \alpha$ is in $\mathcal{S}$. It is also immediate that for each $\psi$ in $\mathcal{S}$,

$$\psi^{-1} * \varphi = \beta$$

is a symmetry of $\mathfrak{X}$ (again because a composition of isomorphisms is an isomorphism). Thus, because

$$\varphi = \psi * \beta,$$

it follows that

$$\varphi * \beta^{-1} = \psi,$$

and thus

$$\alpha \to \varphi * \alpha$$

defines a one-to-one function from the set $G$ of symmetries of $\mathfrak{X}$ onto the scale family $\mathcal{S}$.

The set $G$ of symmetries of $\mathfrak{X}$ together with the operation of functional composition $*$ form a much-investigated, important mathematical structure called a "transformation group":

**Definition 3.1 (group of transformations)** $\mathcal{H} = \langle H, * \rangle$ is said to be a *group of transformations* (or *transformation group*) if and only if for some non-empty set $Y$,

(*i*) the identity function $\iota_Y$ on $Y$ is in $H$;

(*ii*) each element of $H$ is a one-to-one function from $Y$ onto $Y$; and

(*iii*) for all $f$ and $g$ in $H$, $f * g^{-1}$ is in $H$.

By convention, for a group $\mathcal{H} = \langle H, * \rangle$, "$f \in \mathcal{H}$" stands for "$f \in H$," and "$H$ is a group" for "$\mathcal{H}$ is a group." □

**Convention 3.2** Throughout the remainder of this section let

$$\mathcal{G} = \langle G, * \rangle. \quad \square$$

**Definition 3.2** $\mathcal{G}$ is called the *symmetry group of* $\mathfrak{X}$ (or *transformation group of* $\mathfrak{X}$). □

**Theorem 3.1** $\mathcal{G}$ *is a group of transformations.*

**Proof.** Immediate from the definitions of symmetry and isomorphism. □

**Definition 3.3 (homogeneity and uniqueness)** The symmetry group $\mathcal{G} = \langle G, * \rangle$ of $\langle X, \preceq, R_j \rangle_{j \in J}$ is said to be

- *m-point homogeneous* if and only if $m$ is a positive integer and for all $x_1, \ldots, x_m$ and $y_1, \ldots, y_m$ in $X$, if $x_1 \prec \cdots \prec x_m$ and $y_1 \prec \cdots \prec y_m$, then for some $\alpha$ in $G$,

$$\alpha(x_1) = y_1, \ldots, \alpha(x_m) = y_m \,;$$

- *homogeneous* if and only if $\mathcal{G}$ is 1-point homogeneous;

- *n-point unique* if and only if $n$ is a positive integer and for each $\alpha$ and $\beta$ in $G$, if $\alpha$ and $\beta$ agree on some $n$-distinct elements of $X$, then $\alpha = \beta$;

- $\infty$-*point homogeneous* if and only if $\mathcal{G}$ is $m$-point homogeneous for each $m$ in $\mathbb{I}^+$;

- $\infty$-*point unique* if and only if $\mathcal{G}$ is not $n$-point unique for some $n$ in $\mathbb{I}^+$. □

Because of the identification $\alpha \to \varphi * \alpha$ of elements of $G$ with elements of $\mathcal{S}$, all of the relevant information contained in $\mathcal{S}$ for the representational theory of measurement is contained in $\langle G, * \rangle$. ($\mathcal{S}$ contains measurement information not in $\langle G, * \rangle$, for example, the range of its elements—that is, the domain of $\mathfrak{N}$—but such information is not pertinent to the representational theory, because any numerical structure isomorphic to $\mathfrak{X}$ could be selected as the numerical representing structure for $\mathfrak{X}$.) The following is also easy to show for $m$ and $n$ being positive integers or $\infty$:

- $\mathcal{G}$ *is homogeneous if and only if* $\mathcal{S}$ *is homogeneous;*

- $\mathcal{G}$ *is m-point homogeneous if and only if* $\mathcal{S}$ *is m-point homogeneous;*

- *and* $\mathcal{G}$ *is n-point unique if and only if* $\mathcal{S}$ *is n-point unique.*

Using the above, the characterizations discussed in Section 2.5 for the ratio and interval scalability of $\mathfrak{X}$ translate into the following.

**Theorem 3.2** *Suppose* $\mathfrak{X}$ *is a continuous structure. Then the following two statements hold:*

- $\mathfrak{X}$ *is ratio scalable if and only if* $\mathcal{G}$ *is 1-point homogeneous and 1-point unique.*

- $\mathfrak{X}$ *is interval scalable if and only if* $\mathcal{G}$ *is 2-point homogeneous and 2-point unique.*

**Proof.** Narens (1981a,b; also in chapter 2 of Narens, 1985). □

**Definition 3.4** Let $\mathcal{H}$ be a group of transformations and $f$ and $g$ be elements of $\mathcal{H}$. Then $f$ and $g$ are said to *commute* if and only if $f * g = g * f$. $\mathcal{H}$ is said to be *commutative* if and only if $h$ and $k$ commute for all $h$ and $k$ in $\mathcal{H}$.

Because $\mathfrak{X}$ and $\mathfrak{N}$ are isomorphic, they have identical structural properties. As a result, their symmetry groups are isomorphic, and one is commutative if and only if the other is:

**Theorem 3.3** *Let $\mathcal{K} = \langle K, * \rangle$ be the symmetry group of $\mathfrak{N}$. Then the following two statements hold:*

1. *The groups $\mathcal{G}$ and $\mathcal{K}$ are isomorphic.*

2. *$\mathcal{G}$ is commutative if and only if $\mathcal{K}$ is commutative.*

**Proof.** For each $f$ in $G$ and each $r$ in $N$, let
$$\sigma(f)(r) = \varphi * f * \varphi^{-1}(r).$$

Then $\sigma(f)$ is a function from $N$ onto $N$. Because $\varphi$, $f$, and $\varphi^{-1}$ are one-to-one functions, $\sigma(f)$ is a one-to-one function.

Suppose $f$ is an arbitrary element of $G$, $j$ is an arbitrary element of $J$, $n \in \mathbb{I}^+$ is such that the primitive $R_j$ (of $\mathfrak{X}$) is a $n$-ary relation. Then for each $r_1, \ldots, r_n$ in $N$,

$$\begin{array}{ll} S_j(r_1, \ldots, r_n) & \text{iff} \quad R_j(\varphi^{-1}(r_1), \ldots, \varphi^{-1}(r_n)) \\ & \text{iff} \quad R_j(f * \varphi^{-1}(r_1), \ldots, f * \varphi^{-1}(r_n)) \\ & \text{iff} \quad S_j(\varphi * f * \varphi^{-1}(r_1), \ldots, \varphi * f * \varphi^{-1}(r_n)) \\ & \text{iff} \quad S_j(\sigma(f)(r_1), \ldots, \sigma(f)(r_n)). \end{array}$$

Similarly, for each $r$ and $s$ in $N$,
$$r \leq s \quad \text{iff} \quad \sigma(f)(r) \leq \sigma(f)(s).$$

The above shows that $\sigma(f)$ is a symmetry of $\mathfrak{N}$. Thus it has been shown that $\sigma$ is a function from $G$ into the set of symmetries of $\mathfrak{N}$. Suppose $f$ and $g$ are in $G$ and $f \neq g$. Let $a$ in $X$ be such that $f(a) \neq g(a)$. Then, because $\varphi$ is one-to-one,

$$\begin{aligned} \sigma(f)(\varphi(a)) &= \varphi * f * \varphi^{-1}(\varphi(a)) = \varphi[f(a)] \\ &\neq \varphi[g(a)] = \varphi * g * \varphi^{-1}(\varphi(a)) = \sigma(g)(\varphi(a)), \end{aligned}$$

that is, $\sigma(f)(\varphi(a)) \neq \sigma(g)(\varphi(a))$, and thus $\sigma$ is a one-to-one function.

Let $k$ be an arbitrary element of $K$. By an argument similar to the one at the beginning of the proof, the function $\tau$ on $K$ defined by: for each $x$ in $X$,
$$\tau(k)(x) = \varphi^{-1} * k * \varphi(x),$$

Symmetries and the Erlanger Program

is an element of $G$, and thus

$$\sigma[\tau(k)] = \varphi * (\varphi^{-1} * k * \varphi) * \varphi^{-1} = k,$$

which, because $k$ is an arbitrary element of $K$, shows that $\sigma$ is onto $K$.

The above shows that $\sigma$ is a one-to-one function of $G$ onto $K$. Let $f$ and $g$ be arbitrary elements of $G$. Then

$$\begin{aligned}\sigma(f*g) &= \varphi*(f*g)*\varphi^{-1} = (\varphi*f*(\varphi^{-1}*\varphi)*g*\varphi^{-1}\\ &= (\varphi*f*\varphi^{-1})*(\varphi*g*\varphi^{-1}) = \sigma(f)*\sigma(g),\end{aligned}$$

showing that $\sigma$ is an isomorphism of $\langle G, *\rangle$ onto $\langle K, *\rangle$.

Statement 2: If $\langle G, *\rangle$ is commutative, then, because $\sigma$ is an isomorphism of $\langle G, *\rangle$ onto $\langle K, *\rangle$, it follows that for all $f$ and $g$ in $G$,

$$f*g = g*f \text{ iff } \sigma(f)*\sigma(g) = \sigma(g)*\sigma(f),$$

and thus, because $\sigma$ is onto $K$, $\langle K, *\rangle$ is commutative. A similar argument involving $\sigma^{-1}$ shows that the commutativity of $\langle K, *\rangle$ implies the commutativity of $\langle G, *\rangle$. □

**Lemma 3.1** *Suppose $\mathfrak{X}$ is ratio scalable. Let $\mathfrak{M} = \langle \mathbb{R}^+, \leq, T_j\rangle_{j\in J}$ be a numerical representing structure for $\mathfrak{X}$ and $\mathcal{T}$ be a ratio scale family of isomorphisms of $\mathfrak{X}$ onto $\mathfrak{M}$. Then the set of symmetries of $\mathfrak{M}$ is the set of multiplications by positive reals.*

**Proof.** Let $\psi \in \mathcal{T}$ and $\alpha$ be an arbitrary symmetry of $\mathfrak{M}$. Then it is immediate that $\alpha * \psi$ is in $\mathcal{T}$, and thus, because $\mathcal{T}$ is a ratio scale family, $\alpha * \psi = r \cdot \psi$ for some $r$ in $\mathbb{R}^+$. Let $s$ be an arbitrary element of $\mathbb{R}^+$ and $m_s$ be the function from $\mathbb{R}^+$ onto $\mathbb{R}^+$ that is multiplication by $s$. Then, because $\mathcal{T}$ is a ratio scale, $m_s * \psi$ is in $\mathcal{T}$. Thus, because $m_s * \psi$ and $\psi$ are isomorphisms of $\mathfrak{X}$ onto $\mathfrak{M}$, $m_s = (m_s * \psi) * \psi^{-1}$ is a symmetry of $\mathfrak{M}$. □

The following is a useful characterization of ratio scalable structures.

**Theorem 3.4** *Suppose $\mathfrak{X} = \langle X, \preceq, R_j\rangle_{j\in J}$ is a homogeneous continuous structure. Then the following three statements are equivalent:*

1. *$\mathfrak{X}$ is ratio scalable.*

2. *The symmetries of $\mathfrak{X}$ commute.*

3. *They symmetry group $\mathcal{G} = \langle G, *\rangle$ of $\mathfrak{X}$ is 1-point unique.*

**Proof.** Assume Statement 1. Let $\mathfrak{M} = \langle \mathbb{R}^+, \leq, T_j \rangle_{j \in J}$ be a numerical representing structure and $\mathcal{T}$ be a ratio scale of isomorphisms of $\mathfrak{X}$ onto $\mathfrak{M}$. Then by Lemma 3.1, each symmetry of $\mathfrak{M}$ is multiplication by a positive constant. Such multiplications commute. Thus, because $\mathfrak{X}$ and $\mathfrak{M}$ are isomorphic, it follows Statement 2 of Theorem 3.3 (with $\mathfrak{M}$ in place of $\mathfrak{N}$) that the symmetries of $\mathfrak{X}$ commute, and thus Statement 2 has been shown.

Assume Statement 2. Suppose $f$ and $g$ are elements of $G$ and $a$ is an element of $X$ such that $f(a) = g(a)$. Let $x$ be an arbitrary element of $X$. Because by hypothesis $\mathfrak{X}$ is homogeneous, $G$ is homogeneous, and thus let $h$ in $G$ be such that

$$h(a) = x \,. \tag{3.1}$$

From $f(a) = g(a)$, we obtain

$$h(f(a)) = h(g(a)) \,,$$

which by the commutativity of $\mathcal{G}$ (Statement 2) yields,

$$f(h(a)) = g(h(a)) \,,$$

which by Equation 3.1 yields

$$f(x) = g(x) \,,$$

which, because $x$ is an arbitrary element of $X$, yields $f = g$. Thus Statement 3 has been shown.

Assume Statement 3. Statement 1 follows by Statement 1 of Theorem 3.2. $\square$

## 3.2 ERLANGER PROGRAM

In mathematics, three approaches to geometry have evolved. The first is the *synthetic* approach. This is an axiomatic and deductive approach to geometrical objects. It was developed by ancient geometers and perfected in Euclid's 3rd-century, BC, treatise on geometry. It flourished again in the 19th century, where it was systematically applied to non-euclidean geometries.

The second is the *analytic* approach to geometry, which began as a coordinate and algebraic approach developed by Descartes in the early 17th century. Its current version takes $n$-tuples of real numbers as coordinates and considers geometric objects as functions and sets of equations involving the coordinates. Mathematical operations, including those from analysis and algebra, are allowed to be performed on the functions and sets of equations in order to solve geometric problems. These operations need not

## Symmetries and the Erlanger Program

have intuitive geometric content or significance; only the statement of the geometric problem and its solution are required to have intuitive geometric content or significance. Analytic geometry often permits much simpler solutions to geometric problems than those of synthetic geometry.

Both synthetic and analytic approaches to geometry flourished in 19th-century mathematics. New geometries were discovered, and synthetic and analytic approaches for them were developed. Sometimes axiomatic systems for the same geometry used different primitives. The developments produced the following two theoretical issues:

(1) When do different formulations describe the same geometry?

(2) What does it mean for a relation to belong to a geometry?

Answers to these questions could lead to a systematic classification of geometries and provide for a way to bridge the synthetic and analytic approaches. Klein (1872) proposed answers in what has become known as the *Erlanger Program*. They are based on transformation groups and invariants.

**Definition 3.5** Let $\langle H, * \rangle$ be a transformation group on $Y$ (Definition 3.1) and $R$ a $n$-ary relation, $n \geq 1$, on $Y$. Then $R$ is said to be an *invariant of* $H$ if and only if for all $y_1, \ldots, y_n$ in $Y$ and all $h$ in $H$,

$$R(y_1, \ldots, y_n) \text{ iff } R(h(y_1), \ldots, h(y_n)). \quad \square$$

The concept of "invariant" in Definition 3.5 extends to elements $y$ of $Y$ as follows: $y$ is an *invariant of* $H$ if and only if $h(y) = y$ for all $h$ in $H$. It also extends to higher-order relations on $Y$. For example, for each non-empty subset $Z$ of $Y$, that is, for each 1-ary relation $Z$ of $Y$, and each $h$ in $H$, let

$$h(Z) = \{h(z) \,|\, z \in Z\}.$$

Let $\mathcal{W}$ be a non-empty set of non-empty sets of $Y$. $\mathcal{W}$ is an example of a particular kind of higher-order relation on $Y$, namely a 1-ary relation of 1-ary relations of $Y$. By definition, for each $h$ in $H$ let

$$h(\mathcal{W}) = \{h(Z) \,|\, Z \in \mathcal{W}\}.$$

$\mathcal{W}$ is said to be *invariant under* $H$ if and only if for each $h$ in $H$,

$$h(\mathcal{W}) = \mathcal{W}.$$

As an example, consider $Y$ to be the set of points of the euclidean plane. Then each line in the plane is a non-empty subset of $Y$. Let $\mathcal{L}$ be the set of lines. Then $\mathcal{L}$ correspond to the euclidean concept of "line." It is a non-empty set of non-empty subsets of $Y$. Let $H$ be the group of rotations of points of $Y$ about a particular point of $Y$. Then elements of $H$ take lines into lines, that is, for each $h$ in $H$,

$$h(\mathcal{L}) = \mathcal{L},$$

that is, $\mathcal{L}$ is a "higher-order invariant under $H$."

Higher-order invariants appear throughout mathematics and science. For the applications in Part I of this book, first-order invariants—invariant relations or invariant elements of an non-empty set—are mostly sufficient. The theory of higher-order relations and invariants is developed in Part II.

The Erlanger Program identifies geometries with transformation groups: *Two geometric structures (either synthetic or analytic) are said to specify the same geometry if and only if they have isomorphic transformation groups.*

Synthetic geometries consist of a structure of primitives,

$$\mathfrak{Y} = \langle Y, T_j \rangle_{j \in J},$$

where for each $j$ in $J$, $T_j$ is either an element of $Y$ (a 0-ary relation), or a $n_j$-ary, $n_j \geq 1$, relation on $Y$, or a higher-order relation on $Y$ (which includes a relation of $m$-ary relations, $m \geq 0$, on $Y$, a relation of relations of $m$-ary relations on $Y$, etc.), and axioms about the primitives. The transformation group $\langle H, * \rangle$ of $\mathfrak{Y}$ consists of the set $H$ of transformations on $Y$ that leave the primitives of $\mathfrak{Y}$ invariant. (It is immediate that $\langle H, * \rangle$ is a group.) The axioms about the primitives are used to derive group theoretic properties that specify $H$ up to an isomorphism.[6]

The third approach to geometry is a transformational one based on the Erlanger Program: *A geometry $\mathfrak{Y}$ is a transformation group $\langle H, * \rangle$ on a non-empty set $Y$. The relations, 0-ary, n-ary, higher-order, that are invariant under $H$, belong to $\mathfrak{Y}$; those that are not invariant do not belong to $\mathfrak{Y}$.*

As an example, consider the set of points $\mathcal{E}$ consisting of ordered pairs of real numbers. In terms of cartesian coordinates, the euclidean distance function $d$ between elements $(x, y)$ and $(u, v)$ of $\mathcal{E}$ is defined by

$$d[(x, y), (u, v)] = \sqrt{(x - u)^2 + (y - v)^2},$$

and an orientation for angles is specified so that counterclockwise from the positive abscissa to the positive ordinate determines positive orientation.

---

[6]That is, any two structures satisfying the axioms will have isomorphic symmetry groups.

*Symmetries and the Erlanger Program*

$d$ and this orientation specify an analytic geometry on $\mathcal{E}$. The set $E$ of transformations that leave leave $d$ and the orientation invariant are known to be those generated by translations and rotations. Thus $\langle E, * \rangle$ provides a transformational description of a geometry on $\mathcal{E}$, $\boldsymbol{E}$, traditionally called "euclidean plane geometry." There are a number of synthetic axiomatizations of $\boldsymbol{E}$, sometimes with differing primitives. The symmetry group for the structure of primitives of each of these axiomatizations is isomorphic to $\langle E, * \rangle$.

## 3.3 COMPARISON OF GEOMETRIC AND MEASUREMENT-THEORETIC CONCEPTS

**Theorem 3.5** *Suppose $\mathfrak{X} = \langle X, \preceq, R_j \rangle_{j \in J}$ is a continuous structure, $G$ is the set of symmetries of $\mathfrak{X}$, and $R$ is a n-ary relation on $X$. Then the following three statements are equivalent:*

1. *$R$ is $\mathcal{S}$-representationally meaningful (Definition 2.22) for some scale family of isomorphisms $\mathcal{S}$ onto a numerical representing structure for $\mathfrak{X}$.*

2. *$R$ is invariant under $G$.*

3. *For each scale family of isomorphisms $\mathcal{S}^\star$ from $\mathfrak{X}$ onto a numerical representing structure, $R$ is $\mathcal{S}^\star$-representationally meaningful (Definition 2.22).*

**Proof.** By Theorem 2.7, let $\mathfrak{N} = \langle N, \leq, T_j \rangle_{j \in J}$ be a representing numerical structure for $\mathfrak{X}$, $\mathcal{S}$ be a scale family of isomorphisms from $\mathfrak{X}$ onto $\mathfrak{N}$, and $\varphi \in \mathcal{S}$. Let

$$T = \varphi(R)$$

and

$$\mathfrak{X}' = \langle X, \preceq, R_j, R \rangle_{j \in J} \text{ and } \mathfrak{N}' = \langle N, \leq, T_j, T \rangle_{j \in J}.$$

Then $\varphi$ is an isomorphism of $\mathfrak{X}'$ onto $\mathfrak{N}'$.

Assume Statement 1. By Definition 2.22, $\mathcal{S}$ is a scale family of isomorphisms of $\mathfrak{X}'$ onto $\mathfrak{N}'$. Let $G'$ be the set of symmetries of $\mathfrak{X}'$. Then because $\mathcal{S}$ is both the set of isomorphisms from $\mathfrak{X}$ onto $\mathfrak{N}$ and the set of isomorphisms from $\mathfrak{X}'$ onto $\mathfrak{N}'$, it follows from remarks just after Convention 3.1 that

$$G = \{\psi^{-1} * \eta \,|\, \psi \in \mathcal{S} \text{ and } \eta \in \mathcal{S}\} = G'.$$

Thus, because $R$ is invariant under $G'$, it is invariant under $G$, showing Statement 2.

Assume Statement 2. Then $G$ is the set of symmetries of $\mathfrak{X}'$. Let

$$\mathfrak{N}^\star = \langle N^\star, \leq, T_j^\star \rangle_{j \in J}$$

be an arbitrary numerical representing structure for $\mathfrak{X}$. Let $\mathcal{S}^\star$ be the set of isomorphisms from $\mathfrak{X}$ onto $\mathfrak{N}^\star$ and $\varphi^\star$ be an element of $\mathcal{S}^\star$. Let

$$U = \varphi^\star(R).$$

Then
$$\mathfrak{N}^{\star\prime} = \langle N^\star, \leq, T_j^\star, U \rangle_{j \in J}$$

is a numerical representing structure for $\mathfrak{X}'$ and $\varphi^\star$ is an isomorphism of $\mathfrak{X}'$ onto $\mathfrak{N}^{\star\prime}$. Let $\mathcal{S}^{\star\prime}$ be the set of isomorphisms from $\mathfrak{X}'$ onto $\mathfrak{N}^{\star\prime}$. Then, because $\varphi^\star$ is both an isomorphism from $\mathfrak{X}$ onto $\mathfrak{N}^\star$ and from $\mathfrak{X}'$ onto $\mathfrak{N}^{\star\prime}$, it follows from remarks just after Convention 3.1 that

$$\mathcal{S}^\star = \{\varphi^\star * g \,|\, g \in G\} \quad \text{and} \quad \mathcal{S}^{\star\prime} = \{\varphi^\star * g \,|\, g \in G\},$$

that is, $\mathcal{S}^\star = \mathcal{S}^{\star\prime}$, showing Statement 3.

Assume Statement 3. By Theorem 2.7, there exists a scale family of isomorphisms of $\mathfrak{X}$ onto $\mathfrak{N}$. Then Statement 1 is an immediate consequence of Statement 3. $\square$

The following is a correspondence between some of the basic concepts of the representational theory of measurement and the Erlanger Program for geometry:

| | | |
|---:|:---:|:---|
| qualitative structure | $\longleftrightarrow$ | synthetic geometry |
| numerical structure | $\longleftrightarrow$ | analytical geometrical structure |
| representation (isomorphism) | $\longleftrightarrow$ | isomorphism from a synthetic to an analytic geometry |
| representationally meaningful | $\longleftrightarrow$ | belonging to the geometry under consideration |

The above correspondence holds because the form of the representational theory used is based on isomorphisms. When the representational theory was formalized in Scott & Suppes (1958), they used a weaker form of structure preserving numerical functions called "homomorphisms." Homomorphisms are more general in that they can be many-to-one functions, and can be into the domain of the representing structure, rather than onto the domain. Scott and Suppes saw representation and uniqueness theorems as accounting for how numbers entered into science, and from this point of view, the homomorphism version of the representational theory does a better job than the isomorphism version. Suppes & Zinnes (1963) and Pfanzagl (1968) extended the homomorphism version to include forms of "meaningfulness" appropriate for homomorphisms. However, as pointed out in Narens (1981a), theses forms display certain deficiencies as a "meaningfulness concept" that are not present in the isomorphism version of the representational theory. In situations where each homomorphism is an onto-isomorphism, the approaches to meaningfulness of Suppes & Zinnes (1963), Pfanzagl (1968), and Part 1 of this book are equivalent. As for the Erlanger Program, it was only late in the development of measurement theory that a connection to it was made. The classification of scale types in terms of symmetry groups (Narens, 1981b; Alper, 1987) and the classification of concatenation structures according to scale type (Luce & Narens, 1985) are direct applications of Erlanger Program ideas to measurement theory.

Part II of the book provides a formalization of the Erlanger Program and gives a rationale for its concept of geometric object, or equivalently, its concept of "meaningfulness."

## Chapter 4

# Threshold Measurement

Threshold measurement plays an important role in psychophysics and the history of psychology. It is an *indirect method* of measurement, because observers are only asked to judge or guess if one stimulus is more intense than another. In sensory psychology, it is often employed in the following manner.

The observer is presented two stimuli $x$ and $y$ from physical continuum and is asked to choose which is more intense. She is required to pick either $x$ or $y$; she cannot respond "equally intense," or go on to the next trial without making a choice. Within the experiment, the pair $x,y$ is presented many times, and the percentage of times $x$ is chosen over $y$ is recorded. Fixing $x$ and appropriately varying $y$ allows a stimulus $T(x)$ to estimated so that the probability of $T(x)$ being chosen as being more intense than $x$ is .75. By definition, $T(x)$ is said to be "the .75 threshold for $x$." In this case, the notation "$T_{.75}$" could be used instead of "$T(x)$" to emphasize the probability .75 of $T(x)$ $(= T_{.75}(x))$ being chosen over $x$. "$T_{.75}$" is called a ".75 threshold function" for stimuli. Other threshold functions $T_p$, where $.5 \leq p \leq 1$, are similarly defined. $p = .75$ is often used, because, I suppose, .75 is "halfway between chance (.5) and certainty (1.0)." However, in most cases, this use of "halfway" is theoretically unjustified, and other probabilities between .5 and 1 would do equally well.

There are many additional ways of defining threshold functions, for example, where the probability $p$ in "$T_p$" varies with $x$; that is, $T_{p(x)}(x)$ is the stimulus $z$ such that when presented the pair of stimuli $x$ and $z$ the subject chooses $z$ with probability $p(x)$. There are also nonprobabilistic concepts of threshold. Thus in principle many "equally acceptable" threshold functions exist for each subject for the same stimuli. However, Theorem 4.1 below shows that all threshold functions on continua are isomorphic, and thus have identical structural properties. In particular, from a representational perspective, any two threshold functions on continua have identical measurement-theoretic properties. However, if two threshold functions $T_1$ and $T_2$ are on the same continuum $\langle X, \preceq \rangle$, they may be distinguishable relative to one another; for example, it may be the case that $T_1(x) \prec T_2(x)$ for all stimuli $x$.

The rest of this chapter is organized as follows: Section 4.1 of this chapter provides a representational-theoretic account of a single threshold function on a continuum. Section 4.2 interrelates physical measurement and psychological threshold measurement in the form of Weber's law and Fechner's logarithmic law. Section 4.3 investigates purely psychological formulations of "threshold structure;" and Section 4.4 applies meaningfulness considerations as to what is psychological and what is physical in applications of Weber's law to human judgments of intensity.

## 4.1 CONTINUOUS THRESHOLD STRUCTURES

**Definition 4.1 (continuous threshold structure)** $\langle X, \preceq, T \rangle$ is said to be a *continuous threshold structure* if and only if the following three statements hold:

1. $\langle X, \preceq \rangle$ is a continuum.

2. $T$ is a function from $X$ onto $X$ such that for all $x$ and $y$ in $X$,
$$x \prec y \text{ iff } T(x) \prec T(y).$$

3. For each $x$ in $X$, $x \prec T(x)$.

Let $\mathfrak{X} = \langle X, \preceq, T \rangle$ be a continuous threshold structure. Then $T$ is called the *threshold function* of $\mathfrak{X}$. □

Continuous threshold structures $\langle X, \preceq, T \rangle$ naturally appear in many scientific applications. In psychology $T(x)$ is usually interpreted as a discrimination threshold for a stimulus $x$; that is, "$T(x) \prec y$" is interpreted as, "$y$ is discriminatively more intense than $x$."

**Definition 4.2 (Fechner structure)** $\mathfrak{F} = \langle \mathbb{R}, \leq, F \rangle$ is said to be the *Fechner structure* if and only if $F$ is the function on $\mathbb{R}$ such that for all $x \in \mathbb{R}$, $F(x) = x + 1$. □

**Theorem 4.1 (representation theorem)** *Suppose* $\mathfrak{X} = \langle X, \preceq, T \rangle$ *is a continuous threshold structure. Then $\mathfrak{X}$ is isomorphic to the Fechner structure.*

**Proof.** Theorem 3.4 of Narens (1994). □

## Threshold Measurement

It is easy to show that all the axioms for a continuous threshold structure are necessary for the existence of a representation onto the Fechner structure. Theorem 4.1 shows they are also sufficient.

Many results of this chapter are based on properties the symmetry group of a continuous threshold structure $\mathfrak{X}$. Because the symmetry group of $\mathfrak{X}$ is somewhat complicated, it is convenient to analyze it in terms of the more concrete, numerically based Fechner structure $\mathfrak{F}$. Because by Theorems 4.1 and 3.3 the symmetry group of $\mathfrak{X}$ is isomorphic to the the symmetry group of $\mathfrak{F}$, this method of analysis is justified. Examples of symmetries of the Fechner structure are given below.

**Definition 4.3** ($\iota_r$) For each $r$ in $\mathbb{R}$, let $\iota_r(x)$ be the function $x + r$ on $\mathbb{R}$. A function $f$ on $\mathbb{R}$ is called a *translation by $r$ (of $\mathbb{R}$)* if and only if for some $r$ in $\mathbb{R}$, $f = \iota_r$. □

**Lemma 4.1** $\iota_r$ *is a symmetry of the Fechner structure* $\mathfrak{F} = \langle \mathbb{R}, \leq, F \rangle$.

**Proof.** $F$ is the translation $\iota_1$. Let $r$ be in $\mathbb{R}$. Then for all $x$ and $y$ in $\mathbb{R}^+$,
$$x \leq y \text{ iff } \iota_r(x) \leq \iota_r(y)$$
and
$$\begin{aligned}
\iota_r[F(x)] &= \iota_r(x+1) \\
&= (x+1) + r \\
&= (x+r) + 1 = F[\iota_r(x)],
\end{aligned}$$
and thus $\iota_r$ is a symmetry of $\mathfrak{F}$. □

**Lemma 4.2 (homogeneity)** *Let* $\mathfrak{F} = \langle \mathbb{R}, \leq, F \rangle$ *be the Fechner structure. Then the symmetry group of $\mathfrak{F}$ is 1-point homogeneous.*

**Proof.** Let $x$ and $y$ be arbitrary elements of $\mathbb{R}$ and $r = y - x$. Let $\iota_r$ be the translation by $r$ of $\mathbb{R}$ (Definition 4.3). Then $\iota_r$ is an symmetry of $\mathfrak{F}$, and by the definition of $\iota_r$, $\iota_r(x) = y$. □

Translations by $r$ are just one kind of symmetry of $\mathfrak{F}$. The following definition describes a more general kind of symmetry of $\mathfrak{F}$.

**Definition 4.4** ($\alpha_r$) Let $n$ be an arbitrary integer, $(n, n+1]$ the half-open interval of the reals with endpoints $n$ and $n+1$, and $\alpha$ be an arbitrary strictly increasing function from $(0, 1]$ onto $(0, 1]$. For each $r$ in $\mathbb{R}$, define $\alpha_r$ on $\mathbb{R}$ as follows: For each $x$ in $\mathbb{R}$, if $m$ is an integer such that $x \in (m, m+1]$, let
$$\alpha_r(x) = m + r + \alpha(x - m). \quad \square \tag{4.1}$$

**Lemma 4.3** *Let $\alpha$ an arbitrary strictly increasing function from $(0,1]$ onto $(0,1]$ and $r$ be an arbitrary element of $\mathbb{R}$. Then $\alpha_r$ (Definition 4.4) is a symmetry of the Fechner structure $\mathfrak{F} = \langle \mathbb{R}, \leq, F \rangle$.*

**Proof.** Let $x$ be an arbitrary element of $\mathbb{R}$, and by Definition 4.4, let $m$ be the integer such that such that Equation 4.1 holds. Then

$$\begin{aligned} \alpha_r[F(x)] &= \alpha_r(x+1) = (m+1) + r + \alpha_r[(x+1) - (m+1)] \\ &= (m+1) + r + \alpha_r(x-m) \\ &= (m + r + \alpha_r(x-m)) + 1 = F[\alpha_r(x)]. \end{aligned}$$

Thus to show $\alpha_r$ is a symmetry of the Fechner structure $\mathfrak{F}$, it needs only be shown that for all $x$ and $y$ in $\mathbb{R}$,

$$x \leq y \text{ iff } \alpha_r(x) \leq \alpha_r(y).$$

Suppose $m$ is the integer such that $x \in (m, m+1]$, $y$ is in $\mathbb{R}$, $x \leq y$, and $y \in (n, n+1]$, where $n$ is an integer. If $m = n$, then $\alpha(x-m) \leq \alpha(y-m)$, because by assumption $\alpha$ is strictly increasing on $(0,1]$. Thus

$$\alpha_r(x) = m + r + \alpha(x-m) \leq m + r + \alpha(y-m) = \alpha_r(y).$$

Suppose $m \neq n$. Then $m < n$, because $x \leq y$. Thus, because $\alpha(x-m)$ and $\alpha(y-n)$ are in $(0,1]$, $m + \alpha(x-m) \leq n + \alpha(y-n)$. Therefore,

$$\alpha_r(x) = m + r + \alpha(x-m) \leq n + r + \alpha(y-n) = \alpha_r(y). \qquad \square$$

**Lemma 4.4** *Let $\mathfrak{F}$ be the Fechner structure. Let*

$$H = \{\alpha_r \mid r \in \mathbb{R}\},$$

*where $\alpha$ is a strictly increasing function from $(0,1]$ onto $(0,1]$ and $\alpha_r$ is as defined by Equation 4.1. Then*

- *each translation of $\mathbb{R}$ is in $H$,*
- *and $H \subseteq$ the set of symmetries of $\mathfrak{F}$.*

**Proof.** Lemmas 4.1 and 4.3. $\square$

**Lemma 4.5** *Let $\mathfrak{F} = \langle \mathbb{R}, \leq, F \rangle$ be the Fechner structure. Then the symmetry group of $\mathfrak{F}$ is $\infty$-point unique.*

**Proof.** Let $\beta$ and $\gamma$ be strictly increasing functions from $(0,1]$ onto $(0,1]$ such that $\beta = \gamma$ on $(0, \frac{1}{2}]$ and $\beta \neq \gamma$ on $(\frac{1}{2}, 1]$. Then using the notation in Definition 4.4, $\beta_r$ and $\gamma_r$ are in $H$ (defined in Lemma 4.4) and $\beta_r \neq \gamma_r$ for infinitely many elements of $\mathbb{R}$. By Lemma 4.4, $\beta_r$ and $\gamma_r$ are symmetries of $\mathfrak{F}$. Therefore the symmetry group of $\mathfrak{F}$ is $\infty$-point unique. $\square$

*Threshold Measurement* 51

**Theorem 4.2** *Suppose $\mathfrak{X} = \langle X, \preceq, T \rangle$ is a continuous threshold function. Then the family $\mathcal{S}$ of isomorphisms of $\mathfrak{X}$ onto the Fechner structure $\mathfrak{F} = \langle \mathbb{R}, \leq, F \rangle$ is 1-point homogeneous, but not 2-point homogeneous, and is $\infty$-point unique.*

**Proof.** Let $G$ be the symmetry group of $\mathfrak{F}$. $G$ is 1-point homogeneous by Lemma 4.2. The following shows that $G$ is not 2-point homogeneous:

Suppose $G$ were 2-point homogeneous. A contradiction will be shown. By 2-point homogeneity, let $g$ in $G$ be such that

$$.1 < .2, \ 1 < 4, \ g(.1) = 1, \ \text{and} \ g(.2) = 4.$$

However, because $g$ is a symmetry of $\mathfrak{F}$, from

$$.1 < .2 < 1 = F(0) \ \text{and} \ 4 = g(.2) < g(1),$$

it follows that

$$g(.1) = 1 < g(.2) = 4 < g(1) = g(F(0)) = F(g(0)) < F(g(.1)) = F(1) = 2,$$

that is, $4 < 2$, a contradiction.

$G$ is $\infty$-point unique by Lemma 4.5.

By Theorem 4.1, $\mathcal{S} \neq \varnothing$. Let $\varphi$ be an element of the set $\mathcal{S}$ of isomorphisms of $\mathfrak{X}$ onto $\mathfrak{F}$. It easily follows from the fact that $\varphi$ is an isomorphism that the 1-point, but not 2-point, homogeneity of $G$ implies the 1-point, but not 2-point, homogeneity of $\mathcal{S}$, and that the $\infty$-point uniqueness of $G$ implies the $\infty$-point uniqueness of $\mathcal{S}$. □

Theorem 4.2 shows that measurement of a continuous threshold structure by isomorphisms onto the Fechner structure yields a scale type that is 1-point homogeneous, but not 2- homogeneous, and is $\infty$-point unique. This scale type is not in Stevens' classification.

The following lemma is used in Section 4.4.

**Lemma 4.6** *Let $\mathfrak{F} = \langle \mathbb{R}, \leq, F \rangle$ be the Fechner structure and $\beta(x) = x + r$ be a translation of $\mathbb{R}$ (Definition 4.3) such that $0 < r < 1$. Then $\beta$ is not invariant under the symmetry group of $\mathfrak{F}$.*

**Proof.** Suppose $\beta$ were invariant under the symmetry group of $\mathfrak{F}$. A contradiction will be shown. Let $\gamma$ be the function on $\mathbb{R}$ such that for all $x$ in $\mathbb{R}$ and all $m$ in $\mathbb{I}^+$,

$$\text{if } x \in (m, m+1], \ \text{then} \ \gamma(x) = m + 0 + (x - m)^2.$$

Then by Lemma 4.4, $\gamma$ is a symmetry of $\mathfrak{F}$. Because by hypothesis $\beta$ is invariant under symmetries of $\mathfrak{F}$, it follows that for each $x$ in $(0, 1]$,

$$\gamma(\beta(x)) = \beta(\gamma(x)),$$

that is,
$$(x+r)^2 - (x^2+r) = 0,$$
and thus,
$$2xr + r^2 - r = 0. \tag{4.2}$$
But it is impossible that Equation 4.2 holds for all $x$ in $(0,1]$, because it has only one solution for $x$. □

The above partial characterization of the symmetry group of the Fechner structure is all that is needed for the results of this chapter. Narens (1994) shows the equivalent of the following extension of Lemma 4.4: *The symmetries of the Fechner structure is the set,*

$\{\alpha_r \mid \alpha$ *is a strictly increasing function from* $(0,1]$ *onto* $(0,1]$ *and* $r \in \mathbb{R}^+\},$

*where $\alpha_r$ is as defined in Definition 4.4.*

## 4.2 WEBER'S AND FECHNER'S LAWS

**Definition 4.5** $\langle \mathbb{R}^+, \leq, W_r \rangle$ is said to be a *Weber representing structure with threshold* $1+r$ if and only if $r \in \mathbb{R}^+$ and $W_r$ is the function on $\mathbb{R}^+$ that is multiplication by $1+r$. $\mathfrak{W}$ is said to be a *Weber representing structure* if and only if it is Weber representing structure with threshold $1+r$ for some $r$ in $\mathbb{R}^+$.

Let $\mathfrak{W} = \langle \mathbb{R}^+, \leq, W_r \rangle$ be a Weber representing structure with threshold $1+r$. Then $W_r$ is called the *threshold function* of $\mathfrak{W}$, $1+r$ is called the *modified Weber constant* of $\mathfrak{W}$, and $r$ is called the *Weber constant* of $\mathfrak{W}$. Notice that for all $x$ in $\mathbb{R}^+$,
$$r = \frac{W_r(x) - x}{x}. \quad \square$$

**Theorem 4.3** *Suppose* $\mathfrak{X} = \langle X, \preceq, T \rangle$ *is a continuous threshold structure. Then for each $r$ in $\mathbb{R}^+$, $\mathfrak{X}$ is isomorphic to the Weber representing structure with modified Weber constant $1+r$.*

**Proof.** By Theorem 4.1, let $\varphi$ be an isomorphism of $\mathfrak{X}$ is onto the Fechner structure $\mathfrak{F} = \langle \mathbb{R}, \leq, F \rangle$, where $F(u)$ is the function $u+1$. Let $r$ be an arbitrary element of $\mathbb{R}^+$. For each $u$ in $\mathbb{R}$, let
$$\theta(u) = (1+r)^u.$$
Then for each $x$ and $y$ in $X$,

$x \preceq y$ iff $\varphi(x) \leq \varphi(y)$ iff $\theta[\varphi(x)] \leq \theta[\varphi(y)]$ iff $(\theta * \varphi)(x) \leq (\theta * \varphi)(y)$, and

$$\begin{aligned}(\theta * \varphi)[T(x)] &= \theta[\varphi(T(x))] = \theta[F(\varphi(x))] \\ &= \theta(\varphi(x) + 1) = (1+r)^{\varphi(x)+1} \\ &= (1+r) \cdot (1+r)^{\varphi(x)} = (1+r) \cdot \theta[\varphi(x)] = W_r[(\theta * \varphi)(x)]\,.\end{aligned}$$

Thus $\theta * \varphi$ is an isomorphism of $\mathfrak{X}$ onto the Weber structure $\mathfrak{W}$ with modified Weber constant $1 + r$. $\square$

The following lemma is immediate from Theorem 4.3, its proof, and Lemmas 4.4 and 4.6.

**Lemma 4.7** *Let $\mathfrak{W} = \langle \mathbb{R}^+, \leq, W_r \rangle$ be a Weber representing structure with threshold $1 + r$. For each $s$ in $\mathbb{R}^+$, let $f_s$ be the function on $\mathbb{R}^+$ that is multiplication by $s$. Then the following three statements are true.*

1. *$\mathfrak{W}$ is a continuous threshold structure.*

2. *$f_s$ is a symmetry of $\mathfrak{W}$.*

3. *If $s < 1 + r$, then $f_s$ is not invariant under the symmetries of $\mathfrak{W}$.* $\square$

In the measurement of sensation, it is customary in psychology and physiology to represent the psychological relationships among stimuli in terms of their physical measurements. It should be noted that this practice is a scientific convention and not a necessity brought on by the nature of the psychological phenomena under consideration. Thus in employing this convention, care should be taken to distinguish between those conclusions that result from the psychological phenomena under consideration and those that depend in some manner on the convention.

In the measurement of sensation, the physical stimuli under consideration are measurable by a ratio scale. Throughout this book, it will be assumed that the sensory stimuli come from a continuous extensive structure $\mathfrak{P} = \langle X, \preceq, \oplus \rangle$ of physical stimuli. For most psychophysical cases such a structure $\mathfrak{P}$ exists. This assumption is made to simplify exposition: it is not difficult to show that everything presented in the chapter easily extends to the more general case, which assumes that the stimuli, physical or otherwise, under consideration are measurable by a ratio scale.

The physiologist Ernst Weber (1795-1878) conducted experiments showing that all of the senses whose physical stimuli could be measured precisely on a one-dimensional physical scale obeyed a uniform law. Traditionally, this law has been called "Weber's Law." It figures prominently in the history of psychology. The remainder of this section presents a measurement-theoretic treatment of Weber's Law with remarks about its meaningfulness properties. The presentation is based on ideas and results of Narens (1994).

Each continuous threshold structure is representable by a Weber representing structure (Theorem 4.3), and some in the literature have called such representations "Weber's Law." But Weber's Law is really about a more complicated situation relating threshold and physical measurements.

**Definition 4.6 (Weber's Law)** Suppose $\mathfrak{P} = \langle X, \preceq, \oplus \rangle$ is a continuous extensive structure and $\langle X, \preceq, T \rangle$ is a continuous threshold structure. Then $\langle X, \preceq, \oplus, T \rangle$ is said to satisfy *Weber's Law* if and only if there exists a positive real number $c$ such that for each isomorphism $\varphi$ of $\mathfrak{X}$ onto $\langle \mathbb{R}^+, \leq, + \rangle$ and each $x$ in $X$,
$$\varphi(T(x)) = (1+c)\varphi(x). \quad \square$$

Let $\mathfrak{T} = \langle X, \preceq, T \rangle$ be a continuous threshold structure. The existence of isomorphism $\psi$ from $\mathfrak{T}$ onto a Weber representing structure is not a law; it is just a way of representing or measuring $\mathfrak{T}$. However, when $\langle X, \preceq \rangle$ is a physical continuum and $\varphi$ is a measuring function from a traditional ratio scale used to measure $\langle X, \preceq \rangle$, for example, a ratio scale obtained by measuring an extensive structure of the form $\langle X, \preceq, \oplus \rangle$, and it happens that $\varphi$ is also an isomorphism of $\mathfrak{T}$ onto a Weber structure, then there is a law.

**Theorem 4.4 (Weber's Law: qualitative characterization)** *Suppose $\mathfrak{P} = \langle X, \preceq, \oplus \rangle$ is a continuous extensive structure and $\mathfrak{T} = \langle X, \preceq, T \rangle$ is a continuous threshold structure. Then the following two statements are logically equivalent:*

1. *$\langle X, \preceq, \oplus, T \rangle$ satisfies Weber's Law.*

2. *For all $x$ and $y$ in $X$, $T(x \oplus y) = T(x) \oplus T(y)$.*

**Proof.** By Theorem 2.9, let $\mathcal{S}$ be a ratio scale of isomorphisms of $\mathfrak{P}$ onto $\mathfrak{N} = \langle \mathbb{R}^+, \leq, + \rangle$, and let $\varphi \in \mathcal{S}$. By Lemma 3.1, multiplications by positive reals are the symmetries of $\mathfrak{N}$.

Assume Statement 1. Let $c$ in $\mathbb{R}^+$ be such that $\varphi$ is an isomorphisms of $\mathfrak{T}$ onto the Weber representing structure $\mathfrak{W}$ with threshold $1 + c$. Then $\varphi(T)$ is the function on $\mathbb{R}^+$ that is multiplication by $1 + c$. Because multiplications by positive reals are symmetries of $\mathfrak{N}$, $\varphi(T)$ is a symmetry of $\mathfrak{N}$. By Theorem 3.3, $T = \varphi^{-1}(\varphi(T))$ is a symmetry of $\mathfrak{P}$. Therefore, $T(x \oplus y) = T(x) \oplus T(y)$ for all $x$ and $y$ in $X$.

Assume Statement 2. Because $\mathfrak{T}$ is a continuous threshold structure, $T$ is a $\preceq$-strictly increasing function from $X$ onto $X$. Therefore, by Statement 2, $T$ is a symmetry of $\mathfrak{P}$. It then follows from Theorem 3.3 that $\varphi(T)$ is multiplication by a positive real. Because $\mathfrak{T}$ is a continuous threshold structure, $x \prec T(x)$ for all $x$ in $X$. Thus for all $x$ in $X$, $\varphi(x) < \varphi(T(x)) = \varphi(T)(x)$. Therefore let $c$ in $\mathbb{R}^+$ be such that for all $x$ in $\mathbb{R}^+$, $\varphi(T)(x) = (1+c) \cdot x$.

# Threshold Measurement

It then follows that $\varphi$ is an isomorphism of $\mathfrak{T}$ onto the Weber representing structure $\langle \mathbb{R}^+, \leq, \varphi(T) \rangle$ with threshold function $1+c$. $\square$

Note that Statement 2 of Theorem 4.4 corresponds in usual psychophysical situations to an easily performed experiment: Determine thresholds $T(x)$ and $T(y)$ for stimuli $x$ and $y$ and see if $T(x \oplus y) = T(x) \oplus T(y)$. If it does for a reasonable sampling of stimuli, then Weber's law holds experimentally. If for a particular reasonable choice of stimuli $a$ and $b$, $T(a \oplus b) \neq T(a) \oplus T(b)$, then Weber law fails empirically. This is an example of how the measurement-theoretic approach can sometimes be used to design simple experiments—in this case, $T(x \oplus y) = T(x) \oplus T(y)$ for Weber's law—for testing a mathematical model. Note that this is different than a statistical test of the model, that is, different from testing how well $\varphi(T)$ "fits" Weber's law.

The founding of experimental psychology is usually associated with the 1860 publication of Fechner's *Elemente der Psychophysik*. The following is a derivation Fechner's Logarithmic Law from Weber's Law.

**Theorem 4.5 (Fechner's Logarithmic Law)** *Suppose $\mathfrak{P} = \langle X, \preceq, \oplus \rangle$ is a continuous extensive structure, $\mathfrak{T} = \langle X, \preceq, T \rangle$ is a continuous threshold structure, and $\langle X, \preceq, \oplus, T \rangle$ satisfies Weber's Law with modified Weber constant $1 + c$. Let $\varphi$ be an isomorphism of $\mathfrak{P}$ onto $\langle \mathbb{R}^+, \leq, + \rangle$, and let $\psi$ be an isomorphism of $\mathfrak{T}$ onto the Fechner structure $\langle \mathbb{R}^+, \leq, F \rangle$, where $F$ is the function $x + 1$. Then for each $x$ in $X$,*

$$\psi(x) = \log_{1+c}(\varphi(x)).$$

**Proof.** For each $r$ in $\mathbb{R}^+$, let

$$\theta(r) = (1+c)^r.$$

By the proof of Theorem 4.3, $\theta * \psi$ is an isomorphism of $\mathfrak{T}$ onto the Weber structure $\mathfrak{W}$ with modified Weber constant $1+c$. Thus in particular,

$$\theta * \psi(T(x)) = \varphi(T(x)). \tag{4.3}$$

Then by Equation 4.3, for each $x$ in $X$,

$$(1+c)^{\psi(x)+1} = (1+c)^{\psi(T(x))} = \theta[\psi(T(x))]$$
$$= (\theta * \psi)(T(x)) = \varphi(T(x)) = (1+c) \cdot \varphi(x),$$

that is,

$$(1+c)^{\psi(x)+1} = (1+c) \cdot \varphi(x). \tag{4.4}$$

Taking $\log_{1+c}$ of both sides of Equation 4.4 then yields,

$$\psi(x) + 1 = 1 + \log_{1+c}[\varphi(x)]. \quad \square$$

The algebraic development of Fechner's Logarithmic Law culminating in Theorem 4.5 is very different from Fechner's. Fechner used calculus, and based his derivation on an unsound mathematical argument (see Luce & Edwards, 1958, for a discussion). It, however, contained an insight that when appropriately developed yields a rich theory of thresholds, not only for stimuli from a single dimension, but also stimuli from multidimensions (see Dzhafarov & Colonius, 2001).

## 4.3 THRESHOLD STRUCTURES WITH ONLY PSYCHOLOGICAL PRIMITIVES

Let $\mathfrak{X} = \langle X, \preceq, T \rangle$ be a continuous threshold structure. This section shows how to provide a psychological interpretation for $\preceq$ when $X$ and $T$ are considered psychological primitives. In psychophysics, the set $X$, when viewed from psychology, is a set of psychological stimuli, and when viewed from physics is a set of physical objects. $\langle X, \preceq \rangle$ is usually taken to be a physical continuum, and when viewed this way, $\preceq$ is a physical relationship. In psychophysics, the threshold function $T$ is a psychological relationship determined by the observer's psychological behavior.

Consider the binary psychological relation $\prec^\star$ on $X$ defined by,

$x \prec^\star y$ if and only if the observer's behavior show that $y$ is discriminably more intense than $x$.

An example of $x \prec^\star y$ is that the observer indicates at least 75% of the time that "$y$ is more intense than $x$." Then $\prec^\star$ may be considered as being "psychological" because it depends only on the psychological behavior of the observer. Then the structure $\mathfrak{X}_B = \langle X, \prec^\star \rangle$ is completely psychological in the sense that each of its primitives is psychological. Using the same data set, suppose that $T$ is defined on $X$ by, $y = T(x)$ if and only if the observer indicates 75% of the time that "$y$ is more intense than $x$." Then $T$ too may be considered as "psychological." Further suppose that $\preceq$ is a physical ordering on the stimuli $X$, when $X$ is considered a set of physical objects, and that $\langle X, \preceq, T \rangle$ is a continuous threshold structure. Then the following definition and theorem give criteria for $\preceq$ to have a psychological interpretation in terms of $\prec^\star$ and $T$.

**Definition 4.7 (induced discrimination relation)** Let $\mathfrak{X} = \langle X, \preceq, T \rangle$ be a continuous threshold structure. Define $\prec^\star$ on $X$ as follows: for all $x$ and $y$ in $X$,
$$x \prec^\star y \text{ iff } T(x) \preceq y.$$
Then $\prec^\star$ is called the *induced discrimination relation of* $\mathfrak{X}$.  □

*Threshold Measurement* 57

Let $\prec^*$ be an induced discrimination relation of the continuous threshold structure $\mathfrak{X} = \langle X, \preceq, T \rangle$. The following theorem shows that $\preceq$ is directly definable in terms of $\prec^*$ and $X$. We interpret this result as showing that the physical ordering $\preceq$ of the physical structure $\langle X, \preceq \rangle$ is also a psychological ordering, if $\prec^*$ has an independent definition in terms of the psychological data.

**Theorem 4.6** *Suppose $\mathfrak{X} = \langle X, \preceq, T \rangle$ is a continuous threshold structure and $\prec^*$ is the induced discrimination relation of $\mathfrak{X}$ (Definition 4.7) and $x$ and $y$ are arbitrary elements of $X$. Then the following two statements are equivalent:*

1. $x \preceq y$.

2. *For all $z$ in $X$, if $y \prec^* z$ then $x \prec^* z$.*

**Proof.** Suppose Statement 1. To show Statement 2, suppose $z$ is an arbitrary element of $X$ and $y \prec^* z$. It needs only to be shown that $x \prec^* z$. It follows from Statement 1 that $T(x) \preceq T(y)$. Because also $y \prec^* z$, $T(y) \preceq z$. Because $T(x) \preceq T(y)$, it then follows that $T(x) \preceq z$, and thus that $x \prec^* z$.

Suppose Statement 2. To show Statement 1 by contradiction, suppose $y \prec x$. Then $T(y) \prec T(x)$. Because $\langle X, \preceq \rangle$ is a continuum, choose $z$ in $X$ such that $T(y) \prec z \prec T(x)$. Then $y \prec^* z$ but not $x \prec^* z$, contradicting Statement 2. □

**Definition 4.8 (psychological threshold structure)** $\langle X, \preceq, T \rangle$ is said to be a *continuous psychological threshold structure* if and only if it is a continuous threshold structure and each of its primitives is psychological. □

Theorem 4.6 and previous discussion show that the continuous threshold structures that occur typically in psychophysics are psychological threshold structures.

## 4.4 MEANINGFULNESS CONSIDERATIONS

Let $\mathfrak{T} = \langle X, \preceq, T \rangle$ be a continuous threshold structure and $\mathcal{S}$ be a scale of isomorphisms of $\mathfrak{T}$ onto a Weber structure $\mathfrak{W} = \langle \mathbb{R}^+, \leq, W_c \rangle$ with modified Weber constant $(1 + c)$. Then for all $\varphi$ and $\psi$ in $\mathcal{S}$ and all $x$ in $X$,

$$\varphi[T(x)] = (1+c) \cdot \varphi(x) \text{ and } \psi[T(x)] = (1+c) \cdot \psi(x), \quad (4.5)$$

and thus all isomorphisms in $\mathcal{S}$ yield the same modified Weber constant $1 + c$. In terms of the current notation, Narens (1994) makes the following comments about this:

Some measurement theorists might want to use this result to say that "$1 + c$ is meaningful." I think this would be a error: This *by itself* is not enough to conclude that "$1 + c$ is meaningful;" it is only enough to conclude that the sentence

$$\varphi[T(x)] = (1 + c) \cdot \varphi(x)$$

is a meaningful assertion. To properly conclude "$1+c$ is meaningful," additional observations like the following are needed:

> Multiplication by the constant $1 + c$ is an symmetry of $\mathfrak{W}$, and it is [invariant under the symmetries of $\mathfrak{W}$] since it is the threshold function $W_c$. Through the isomorphisms of $\mathcal{S}$, it has an interpretation in $\mathfrak{T}$ as the threshold function $T$.

Observations similar to the above do not hold for the Weber constant $c$. Consider the Weber formula

$$\varphi[T(x)] - \varphi(x) = c \cdot \varphi(x), \qquad (4.6)$$

where $\varphi \in \mathcal{S}$. By [Lemma 4.7], multiplication by $c$ is not [invariant under the symmetries of $\mathfrak{W}$]. By using results of Narens (1988) [which are similar to results in Part II of this book], this means that the symmetry of $\mathfrak{T}$ that corresponds to multiplication by $c$ via $\varphi^{-1}$ is not definable in terms of the primitives $X$, $\preceq$, and $T$, no matter how powerful a logical language is used.

Note that the statement in Equation 4.6 is meaningful in the sense that if $\psi$ is any element of $\mathcal{S}$, then

$$\psi[T(x)] - \psi(x) = c \cdot \psi(x).$$

The fact that this statement is "meaningful" does not mean that every part of it—e.g., the constant $c$—has a proper interpretation in $\mathfrak{T}$.

Let $\mathfrak{P} = \langle X, \preceq, \oplus \rangle$ be a continuous extensive structure and $\mathfrak{T} = \langle X, \preceq, T \rangle$ be a continuous threshold structure. For this discussion, $X$ will be considered as a set of physical objects as well as a set of psychological stimuli, $\preceq$ will be considered as a physical relation on physical objects as well as a psychological relation on psychological stimuli, $\oplus$ will be considered as a physical operation on physical objects, and $T$ will be considered as a psychological function on psychological stimuli. Thus $\mathfrak{P}$ characterizes a physical situation and $\mathfrak{T}$ characterizes a psychological situation. Let $\mathcal{S}$ be a ratio scale of isomorphisms of $\mathfrak{P}$ onto

# Threshold Measurement 59

$\langle \mathbb{R}^+, \leq, + \rangle$. As discussed above, the modified Weber constant that results from measurement by $\mathcal{S}$ has an interpretation in the psychological structure $\mathfrak{T}$, whereas the Weber constant has no such interpretation. By using results of Narens (1988) [which are similar to results in Part II of this book], both constants have interpretations in the physical structure $\mathfrak{P}$. (Here "interpretability" means definable from the primitives of $\mathfrak{P}$ in term of some sufficiently powerful logical language.) (pp. 320–321)

Weber constants are used in many ways in psychology. One is as a measurement of psychological sensitivity. For example, individuals are ranked according to how sensitive they are to a physical continuum of stimuli by Weber constants of their thresholds for the stimuli, the smaller the Weber constant, the more sensitive to the stimuli. Although the Weber constants are in themselves not meaningful (in the sense of the previous passage of Narens, 1994, in that they are not interpretable as part of the psychology inherent in their thresholds), their use in the above method of ranking is still proper. This is because the same ranking results by ordering the subjects by their modified Weber constants. This case is special, because for Weber constants $c$ and $d$, $c < d$ iff $1 + c < 1 + d$. There are many other situations involving psychological sensitivity where the use of Weber constants produce an end result that is not interpretable as part of the psychology inherent in the thresholds. (See Narens & Mausfeld, 1992, for examples and a discussion regarding the relationship of the psychological and the physical in psychophysics.)

It should be emphasized that "meaningfulness" throughout this book is taken as a relative and not as an absolute concept. Consider the Weber Law structure $\mathfrak{W} = \langle X, \preceq, \oplus, T \rangle$, where $X$, $\preceq$, and $T$ are psychological and $\oplus$ is not psychological. Then the Weber constant $c$ associated with $\mathfrak{W}$ is not psychologically meaningful for $\mathfrak{T} = \langle X, \preceq, T \rangle$, because there is not enough psychology in $\mathfrak{T}$ to appropriate express $c$. Suppose additional psychological primitives are added to those of $\mathfrak{T}$ to produce a structure $\mathfrak{Y}$ that is measurable by a ratio scale $\mathcal{S}$ of isomorphisms. Then $\mathfrak{Y}$ has enough psychology to make $c$ psychologically meaningful by most of the concepts of "meaningfulness" discussed in this book, for example, $\mathcal{S}$-representational meaningfulness.

Chapter 5

# Magnitude Production

## 5.1 STEVENS' METHODS OF MAGNITUDE ESTIMATION AND PRODUCTION

Magnitude estimation and production are methods of direct measurement developed and promoted the psychologist S. S. Stevens during the midpart of the 20th century. They are widely used in the social and behavioral sciences. Stevens developed these methods together with a theory of measurement as a response to a widely held view of the time that justifiable forms of measurement that were stronger than counting or numerical ordering necessarily relied on the existence of an observable, empirical form of addition, for example, measurement through extensive structures. Because psychological phenomena generally lacked such forms of addition, many scientists of the time believed that psychology was not—and could never be—a quantitative science founded on philosophically sound principles. Stevens and other psychologist disagreed. This led to the British Association for the Advancement of Science to appoint a committee to look into the matter. Stevens (1946) comments,

> For seven years a committee of the British Association for the Advancement of Science debated the problem of measurement. Appointed in 1932 to represent Section A (Mathematical and Physical Sciences) and Section J (Psychology), the committee was instructed to consider and report upon the possibility of "quantitative estimates of sensory events"—meaning simply: Is it possible to measure human sensation? Deliberation led only to disagreement, mainly about what is meant about the term measurement. An interim report in 1938 found one member complaining that his colleagues "came out by the same door as they went in," and in order to have another try at agreement,

the committee begged to be continued for another year.

For its final report (1940) the committee chose a common bone for its contentions, directing its arguments at a concrete example of a sensory scale. This was the Sone scale of loudness (S. S. Stevens & H. Davis. *Hearing.* New York: Wiley, 1938), which purports to measure the subjective magnitude of an auditory sensation against a scale having the formal properties of basic scales, such as those used to measure length and weight. Again 19 members of the committee came out by the routes they entered, and their views ranged widely between two extremes. One member submitted "that any law purporting to express a quantitative relation between sensation intensity and stimulus intensity is not merely false but is in fact meaningless unless and until a meaning can be given to the concept of addition as applied to sensation" (Final Report, pg. 245). (pg. 667)

In this chapter, I present what I consider to be the soundest version of the ideas inherent in Steven's methods of direct measurement. Although articulately presented, Stevens' writings lacked the necessary mathematical and philosophical rigor to present clearly his insights. This undoubtedly contributed to making his highly controversial methods of measurement more controversial than necessary—that is, his methods would have been less controversial if given a careful, consistent, formal presentation.[7]

**Convention 5.1** In discussing magnitude estimation, it is important to distinguish number words that name numbers from numbers. Throughout this book, the convention of putting number words in boldface type is often adopted. Thus **3** is a number word denoting the number 3. In magnitude production, this convention is often employed as follows: The ordered triple, $(x, \mathbf{3}, t)$ stands for *the observer when instructed to produce a stimulus that is 3 times t, chooses x.* □

In *ratio magnitude production,* the observer is presented with a stimulus $t$ from a physical dimension of stimuli $X$ and is asked to produce a stimulus $x$ that is "$p$ times $t$," where $p$ is some positive number, usually an integer or a fraction. This yields the data $(x, \boldsymbol{p}, t)$ that the experimenter then uses for constructing a scale family of measuring functions.

---

[7]There were earlier measurement-based approaches to magnitude production, estimation, and the related technique of cross modality matching, than the ones of Narens (1996, 1997) that this chapter is based on. (For example, Shepard, 1978, 1981; Krantz, 1972; Marley, 1972; Miyamoto, 1983; and Luce, 1990. Krantz's, Luce's, and Miyamoto's theories are axiomatic; Marley's is probabilistic.) Narens' approach is fundamentally different from these earlier ones.

*Magnitude Production*

In *ratio magnitude estimation,* the observer is presented two stimuli $x$ and $t$ and is asked to name the ratio of his numerical estimate of the subjective intensity of $x$ to his numerical estimate of the subjective intensity of $t$. If the observer says $\boldsymbol{p}$, this yields data $(x, \boldsymbol{p}, t)$ that the experimenter uses for constructing a scale family of measuring functions.

Because ratio magnitude production yields data sets having the same form as those from ratio magnitude estimation, the theory developed in this chapter for ratio magnitude production can be modified so that it applies to ratio magnitude estimation.

## 5.2 NARENS' 1996 THEORY

Narens (1996) isolates two principles that he believes are inherent in Stevens' direct methods of measurement. These principles are not stated by Stevens, but appear to be implied by ($i$) his data collection methods, ($ii$) his methods of statistical analysis, and ($iii$) the conclusions he drew from his experimental research. The following is a reformulation of Narens (1996):

> **Principle 1.** $\mathcal{S}$ is a ratio scale family that adequately measures the observer's subjective intensity of stimuli in $X$.
>
> **Principle 2.** For each $t$ in $X$, there is a function $\varphi_t$ in $\mathcal{S}$ such that (1) $\varphi_t(t) = 1$, and (2) for each $x$ in $X$, if the observer's behavior yields the data,
>
> $$(x, \boldsymbol{p}, t),$$
>
> then $\varphi_t(x) = p$.

Narens shows (see Theorem 5.1 below) that Principles 1 and 2 imply the following property.

**Definition 5.1 (Multiplicative Property)** Using Convention 5.1, the *Multiplicative Property* is said to hold for $p$, $q$, and $r$ in $\mathbb{R}^+$ if and only if

if $(x, \mathbf{p}, t)$, $(y, \mathbf{q}, x)$, and $(y, \mathbf{r}, t)$, then $r = q \cdot p$.  $\square$

**Theorem 5.1** *Suppose $\mathcal{S}$ is a scale family from $X$ onto $\mathbb{R}^+$ that satisfies Principles 1 and 2 above. Then the Multiplicative Property (Definition 5.1) holds for all $p$, $q$, and $r$ in $\mathbb{R}^+$.*

**Proof.** Let $t \in X$, and $p$, $q$, and $r$ be arbitrary elements of $\mathbb{R}^+$. Suppose $(x, \mathbf{p}, t)$, $(y, \mathbf{q}, x)$, and $(y, \mathbf{r}, t)$. Then $\varphi_t(x) = p$, $\varphi_x(y) = q$, and $\varphi_t(y) = r$. By Principle 1, let $u$ in $\mathbb{R}^+$ be such that

$$\varphi_x = u\varphi_t. \tag{5.1}$$

Then by Principle 2, $1 = \varphi_x(x) = u\psi_t(x) = u \cdot p$, that is,

$$u = \frac{1}{p}. \tag{5.2}$$

Therefore, by Equations 5.1 and 5.2,

$$q = \varphi_x(y) = u \cdot \varphi_t(y) = \frac{1}{p} \cdot \varphi_t(y) = \frac{1}{p} \cdot r,$$

that is, $r = q \cdot p$. □

Narens (1996) comments,

> In almost all the cases in the literature either insufficient or the wrong kind of data have been collected for the valid testing of the multiplicative property. However, because of the strong structural relationship it describes, I suspect that the multiplicative property would fail empirically for most of the kinds of situations where magnitude estimation is employed. If such massive failures are indeed the case, then in light of Theorem 5.1 something fundamental must be changed in the preceding theory of ratio magnitude estimation.
>
> In my opinion, it is Stevens' method of constructing the representations $\varphi_t$ that is most suspect: I see no reason why just because an observer says or indicates for a fix modulus $t$ and for various $x$ in $X$ that "$x$ is $p$ times more intense than $t$," it then follows that $\varphi_t(x) = p$ is a valid representation of the observer's subjective intensity of $x$ with respect to $t$. Stevens and other magnitude estimation theorists do not provide any theoretical or even intuitive rationale for this; at most they only note that the method of ratio magnitude estimation yields representations (that they presume to be parts of ratio scales) that interrelate in consistent and theoretically interesting ways with other phenomena. It is my conjecture that the consistency results not by reflecting some underlying reality but from a lack of enough relevant data that might reveal structural inconsistencies. (pg. 110)

Narens (1996) provides axiomatizations of magnitude production behavior. These axiomatizations were designed to illustrate the role of the Multiplicative Property as the driving force behind Stevens' method of measurement for magnitude production. They include the following three axioms.

# Magnitude Production

**Axiom 5.1** $\langle X, \preceq \rangle$ is a continuum. □

**Axiom 5.2** The following five statements are true:

1. For all $p$ in $\mathbb{I}^+$ and all $t$ in $X$ there exist $x$ in $X$ such that $(x, \mathbf{p}, t)$ is in the data.

2. For all $x$ and $t$ in $X$ and all $p$ in $\mathbb{I}^+$, if $(x, \mathbf{p}, t)$ is in the data, then $t \preceq x$.

3. For all $t$ in $X$, $(t, \mathbf{1}, t)$ is in the data.

4. For all $x$ and $t$ in $X$ and all $p$ in $\mathbb{I}^+$, there exist exactly one $z$ in $X$ and exactly one $s$ in $X$ such that $(z, \mathbf{p}, t)$ and $(x, \mathbf{p}, s)$ are in the data.

5. For all $x$, $y$, $t$, and $s$ in $X$ and $p$ in $\mathbb{I}^+$, if $(x, \mathbf{p}, t)$ and $(y, \mathbf{p}, s)$ are in the data, then
$$x \preceq y \text{ iff } t \preceq s. \quad \square$$

Statements 1 to 5 of Axiom 5.2 are straightforward and are natural generalizations for empirical ratio magnitude production phenomena in many domains.[8]

**Axiom 5.3** The following three statements are true:

1. For all $(x, \mathbf{p}, t)$ and $(y, \mathbf{q}, t)$ in the data,
$$x \prec y \text{ iff } p < q.$$

2. For all $x$ and $t$ in $X$, if $t \prec x$, then there exist $y$ in $X$ and $p$ in $\mathbb{I}^+$ such that $x \prec y$ and $(y, \mathbf{p}, t)$ is in the data.

3. For all $x$ and $t$ in $X$, if $t \prec x$, then there exist $y$ and $z$ in $X$ and $p$ in $\mathbb{I}^+$ such that
$$(y, \mathbf{p+1}, t) \text{ and } (y, \mathbf{p}, z)$$
are in the data and
$$t \prec z \prec x. \quad \square$$

---

[8]Statement 3 may be problematic in some situations where the stimuli are presented sequentially and the subjective intensity of the second stimulus is influenced by the first. Such sequential presentations occur in audition experiments. In vision experiments, the stimuli can be presented either simultaneously or sequentially.

Axiom 5.3 describes natural conditions for a ratio magnitude production paradigm. Statement 1 provides the linkage of the usual ordering on numbers, and consequently the usual ordering on number words to the experimenter determined ordering on stimuli. Statement 2 is an "Archimedean axiom" which guarantees that no element of $X$ is "infinitely large" in terms of ratio magnitude production with respect to another element of $X$. Statement 3 is also an "Archimedean axiom" that essentially says that no two distinct elements are "infinitesimally close" in terms of ratio magnitude production. In terms of real numbers, the Statement 3 is similar in concept to the following: For all positive reals $t'$ and $x'$, if $t' < x'$, then for some $p$ in $\mathbb{I}^+$ and $z$ in $\mathbb{R}^+$,

$$t' < z = \frac{p+1}{p}t' < x'.$$

**Definition 5.2** Assume the behavioral assumptions Axioms 5.1 to 5.3. For each $p$ in $\mathbb{I}^+$, define the binary relation $\overline{p}$ on $X$ as follows: For all $x$ and $t$ in $X$,

$$x = \overline{p}(t) \text{ iff } (x, \mathbf{p}, t) \text{ is in the data.}$$

$\overline{p}$ is called the *behavioral interpretation* of $\mathbf{p}$. □

Because $\overline{p}$ is defined entirely in terms of behavioral concepts, it is also a behavioral concept.

It easily follows from Axioms 5.1 to 5.3 that for each $p$ in $\mathbb{I}^+$, $\overline{p}$ is a function on $X$.

**Definition 5.3** Let

$$\mathfrak{D} = \langle X, \preceq, \overline{1}, \ldots, \overline{p}, \ldots \rangle_{p \in \mathbb{I}^+}.$$

By definition, $\mathfrak{D}$ is called the *behavioral structure (associated with the data)*.

□

**Convention 5.2** Throughout the rest of this section, $\mathfrak{D}$ denotes the behavioral structure associated the collected magnitude production data. □

Note that each primitive of $\mathfrak{D}$ is a behavioral concept.

**Definition 5.4** $\varphi$ is said to be a *Stevens' measuring function* for $\mathfrak{D}$ if and only if $\varphi$ is a function from $X$ into $\mathbb{R}^+$ such that for each $p \in \mathbb{I}^+$, $\varphi(\overline{p})$ is the function that is multiplication by $p$.

A scale family $\mathcal{S}$ on $X$ is said to be a *Stevens scale* for $\mathfrak{D}$ if and only if (i) each element of $\mathcal{S}$ is a Stevens' measuring function for $\mathfrak{D}$, and (ii) $\mathcal{S} = $ the set of isomorphisms of $\mathfrak{D}$ onto $\langle \mathbb{R}^+, \leq, \varphi(\overline{p}) \rangle_{p \in \mathbb{I}^+}$.

## Magnitude Production

More generally, $\varphi$ is said to be a *multiplicative measuring function* for $\mathfrak{D}$ if and only if $\varphi$ is a function from $X$ into $\mathbb{R}^+$ such that for each $p \in \mathbb{I}^+$, $\varphi(\overline{p})$ is a function that is multiplication by a positive real. And a scale family $\mathcal{S}$ on $X$ is said to be a *multiplicative scale* for $\mathfrak{D}$ if and only if each element of $\mathcal{S}$ is a multiplicative measuring function for $\mathfrak{D}$. $\square$

Suppose

- $(x, \mathbf{p}, t)$ is in the data set, that is, the observer selects $x$ to be "$p$ times" $t$,
- $\varphi$ is a Stevens' measuring function for $\mathfrak{D}$,
- and $\psi$ is a multiplicative measuring function for $\mathfrak{D}$.

Then
$$\varphi(x) = \varphi(\overline{p})(\varphi(t)) = p \cdot \varphi(t) \,.$$
Let $f(p)$ be the positive real such that $\psi(\overline{p})$ is multiplication by $f(p)$. Then
$$\psi(x) = \psi(\overline{p})(\psi(t)) = f(p) \cdot \psi(t) \,.$$
Thus $\varphi$ veridically interprets the instruction, "Select $x$ that is $p$ times $t$," because $x$ and $t$ are in a ratio relationship and that this ratio is $p$. $\psi$ only partially veridically interprets the instruction, "Select $x$ that is $p$ times $t$": It veridically interprets $x$ and $t$ in being in a ratio relationship, but does not necessarily veridically interprets this ratio, $f(p)$, as being $p$.

In the context of Axioms 5.1 to 5.3, the following axiom is necessary and sufficient for $\mathfrak{D}$ to have a multiplicative scale family of isomorphisms.

**Axiom 5.4 (Commutative Axiom)** For all $p$ and $q$ in $\mathbb{I}^+$ and all $x$, $y$, $z$, $t$, and $w$ in $X$, if $(x, \mathbf{p}, t)$, $(z, \mathbf{q}, x)$, $(y, \mathbf{q}, t)$, and $(w, \mathbf{p}, y)$ are in the data, then $z = w$. $\square$

Let $\mathbf{q} \bullet \mathbf{p}$ stand for first estimating $p$ times a stimulus $t$ and then $q$ times that estimated stimulus. Then Axiom 5.4 says that $\mathbf{q} \bullet \mathbf{p} = \mathbf{p} \bullet \mathbf{q}$.

**Theorem 5.2** *Assume Axioms 5.1 to 5.4. Then the following two statements are true:*

1. *There exists a numerical structure $\mathfrak{N}$ such that the set of isomorphisms of $\mathfrak{D}$ onto $\mathfrak{N}$ is both a ratio scale and a multiplicative scale for $\mathfrak{D}$.*

2. *Suppose $\mathcal{S}$ is a ratio scale of isomorphisms of $\mathfrak{D}$ onto a numerical structure. Then $\mathcal{S}$ is a multiplicative scale for $\mathfrak{D}$.*

**Proof.** Narens (1996). □

In the context of Axioms 5.1 to 5.3, the following axiom is necessary and sufficient for $\mathfrak{D}$ to have a Stevens' scale family of isomorphisms:

**Axiom 5.5 (Multiplicative Axiom)** For all $p$ and $q$ in $\mathbb{I}^+$ and all $x$, $y$, $z$, and $t$ in $X$, if $(x, \mathbf{p}, t)$, $(z, \mathbf{q}, x)$, and $(z, \mathbf{r}, t)$, then $r = q \cdot p$. □

**Theorem 5.3** *Assume Axioms 5.1 to 5.3 and 5.5. Then there exists a numerical structure $\mathfrak{N}$ such that the set of isomorphisms of $\mathfrak{D}$ onto $\mathfrak{N}$ is both a ratio scale and a Stevens' scale for $\mathfrak{D}$.*
**Proof.** Narens (1996). □

It is easy to reformulate Axioms 5.1 to 5.5 so that they apply to fraction names of the form $\frac{1}{n}$, $n \in \mathbb{I}^+$, and show the obvious reformulations of Theorems 5.2 and 5.3 for such fractions.

## 5.3 EMPIRICAL TESTS

Narens (1996) suggested that the Multiplicative Axiom would likely fail empirically and implicitly suggested using the Commutative Axiom (Axiom 5.4) as a test for the ratio scalability magnitude production data. Several researchers have tested the two axioms in psychophysical production experiments:

Ellermeier & Faulhammer (2000) tested the Multiplicative and Commutative Axioms in a standard loudness magnitude production paradigm. They found that the Multiplicative Axiom fails and the Commutative Axiom holds. For example, they found **2•3 ≠ 6** but **2•3** approximately equals **3 • 2**. Results of Ellermeier & Faulhammer indicated that for the stimuli they tested, **2 • 3** was generally much larger than **6** in the sense that if a stimulus $x$ was produced that was three times as loud a stimulus $t$ followed later by the production of a stimulus $y$ that was twice as loud as $x$, then $y$ was about the production of twelve times $t$, that is, **2 • 3** approximately equals **12**.

Peißner (1999) performed a vision study that paralleled the design of Ellermeier & Faluhammer. He obtained similar results regarding the failure of the Multiplicative Axiom and the holding of the Commutative Axiom.

Steingrimsson & Luce (2005) also observed the holding of the Commutativity Axiom in a loudness experiment for instructions producing stimuli that were two times and three times other stimuli.

Zimmer (2005) also found Multiplicative Axiom to fail in a loudness production experiment but the Commutative Axiom to hold. Zimmer tested stimuli that were one half as loud, one third as loud, and one sixth as loud as other stimuli.

Magnitude Production

## 5.4 CONTINUOUS RATIO PRODUCTION

The theory of magnitude production of the previous section was designed to isolate assumptions inherent in Stevens' theory. One of these was the Multiplicative Axiom (Axiom 5.5), which, as previously discussed, fails empirically. Another assumption that fails empirically in some loudness experiments is Condition 3 of Axiom 5.2, "For all $t$ in $X$, $(t, \mathbf{1}, t)$ is in the data." It fails, because in loudness experiments stimuli are presented sequentially, and the loudness of the first stimulus may influence the loudness of the second. Both of these failures result from number words in the instructions being given veridical interpretations. The following simple axiomatization of a ratio production paradigm does not use number words.

**Definition 5.5 (continuous ratio production)** $\langle X, \preceq, R_{ab}\rangle_{a,b \in X}$ is said to be a *continuous ratio production structure* if and only if the following four axioms hold:

1. **Continuum of Stimuli:** $\langle X, \preceq \rangle$ is a continuum of stimuli.

2. **Ratio Production:** To each pair of stimuli $a$ and $b$ and for each stimulus $t$ in $X$, the observer can adjust a stimuli $x$ to satisfy the command, "Find a stimulus $x$ so that the ratio of the loudness of $a$ to $b$ is the same as the ratio of the loudness of $t$ to $x$." These adjustments produce a $\preceq$-strictly increasing function $R_{ab}$ on $\langle X, \preceq \rangle$ defined by,
$$R_{ab}(t) = x \,.$$

3. **Homogeneity:** For each $t$ and $x$ in $X$, there exist $a$ and $b$ in $X$ such that
$$R_{ab}(t) = x \,.$$

4. **Commutativity:** For all $a$, $b$, $c$, and $d$,
$$R_{ab} * R_{cd} = R_{cd} * R_{ab} \,. \quad \square$$

In the axiomatization of Continuous Ratio Production, the axiom of Continuum of Stimuli and variants of the axiom of Ratio Production correspond to assumptions routinely made in many ratio production paradigms. The axiom of Homogeneity may be viewed as a continuity principle; and the axiom of Commutativity corresponds to a simple experiment.

**Theorem 5.4** *Suppose* $\mathfrak{X} = \langle X, \preceq, R_{ab} \rangle_{a,b \in X}$ *is a continuous ratio production structure. Then there exist functions* $S_{ab}$ *on* $\mathbb{R}^+$ *and a family of* functions $\mathcal{S}$ such that $\mathcal{S}$ is the ratio scale family of isomorphisms of $\mathfrak{X}$ onto $\mathfrak{N} = \langle \mathbb{R}^+, \leq, S_{ab} \rangle_{a,b \in X}$.

**Proof.** Let $a$ and $b$ be arbitrary elements of $X$. Then by the axiom of Ratio Production, $R_{ab}$ is $\preceq$-strictly increasing and thus for all $u$ and $v$ in $X$,

$$u \preceq v \text{ iff } R_{ab}(u) \preceq R_{ab}(v). \tag{5.3}$$

Let $c$ and $d$ be arbitrary elements of $X$. By the axiom of Commutativity, for each $u$ in $X$,

$$R_{ab}[R_{cd}(u)] = R_{cd}[R_{ab}(u)]. \tag{5.4}$$

By Equations 5.3 and 5.4, $R_{ab}$ is a symmetry of $\mathfrak{X}$. Let $G$ be the symmetry group of $\mathfrak{X}$. Then by the axiom of Homogeneity, $G$ is 1-point homogeneous.

To show $G$ is 1-point unique, let $\alpha$ be and $\beta$ arbitrary elements of $G$, and $x$ in $X$ be such that $\alpha(x) = \beta(x)$. Let $y$ be an arbitrary element of $X$. By the axiom of Homogeneity, let $g$ and $h$ be elements of $X$ such that

$$R_{gh}(x) = y.$$

Then, because $R_{gh}$ is a primitive of $\mathfrak{X}$ and $\alpha$ and $\beta$ are symmetries of $\mathfrak{X}$,

$$\begin{aligned}\alpha(y) &= \alpha[R_{gh}(x)] = R_{gh}[\alpha(x)] \\ &= R_{gh}[\beta(x)] = \beta[R_{gh}(x)] = \beta(y),\end{aligned}$$

that is, $\alpha(y) = \beta(y)$. Because $y$ was chosen to be an arbitrary element of $X$, $\alpha = \beta$.

Because $G$ is 1-point homogeneous and 1-point unique, it follows by Theorem 3.2 that there exists $\mathfrak{N} = \langle \mathbb{R}^+, \leq, S_{ab} \rangle_{a,b \in X}$ and $\mathcal{S}$ such that $\mathcal{S}$ is a ratio scale of isomorphisms of $\mathfrak{X}$ onto $\mathfrak{N}$. $\square$

## 5.5 CONCLUSIONS

Stevens and many others in the behavioral sciences have used direct measurement methods to produce scale families. The theory and empirical results of this chapter suggest that such methods are highly suspect. However, the *data* yielded by direct measurement elicitation procedures may still be informative. This is particularly the case if it can be established that the empirical relationships on stimuli that correspond to instructions to produce or estimate individual ratios are functions that commute with one another. In such cases, the data is consistent with the *indirect measurement* of the stimuli via a ratio scale of isomorphisms that represents each such empirical relationship by a numerical ratio—one generally different from the ratio named in the instructions that gave rise to the empirical relationship.

It should be noted that, in the indirect approach, it is not assumed that the observer has a scientific understanding of numbers or ratios or even uses analogs of numbers or ratios in her calculations that yield the produced stimuli. The ratios that appear in the numerical representations result from the experimenter's choice of a particular representing structure. According to the representational theory, other representing structures are equally valid, and most would not represent the empirically based behavioral functions on stimuli as ratios. The convention of this chapter of representing such behavioral functions as ratios is not based on science, but may be viewed as a homage to Stevens and other pioneers of direct measurement.

# Chapter 6

# Torgerson's Conjecture

## 6.1 BISECTION DATA

Plateau's theory of the shape of the psychophysical function was described in chapter 1. It was based, in part, on empirical observations about artists' productions of "midway" grays painted on disks. Plateau assumed that the artists used subjective equal ratio judgments in painting their grays. However, because he only asked the artists to find the "midway" gray, it is ambiguous from the perspective of direct measurement, whether a given artist used equal ratios or equal differences or some other interpretation of "midway." As discussed at the end of Section 1.1, Plateau was aware of the potential ambiguity, for he wrote,

> it seemed to me more rational, in order to explain the invariance of the general effect of the picture, to postulate a priori the constancy of the ratios and not the differences.

In terms of direct measurement methodology, Plateau should have been more explicit in his instructions, for example, instructing the artists to employ equal ratios in determining their midway grays. However, according to the representational theory, such instructions would not resolve the ambiguity of whether to represent the bisection data by the geometric mean (equal ratios) or by the arithmetic mean (equal differences): For according to the representational theory, if the data are consistent with the properties of a bisection structure (Definition 2.14), then both forms of representation are valid.

Let $\mathfrak{X} = \langle X, \preceq, \ominus \rangle$ be a bisection structure (Definition 2.14). By Theorem 2.8, $\ominus$ has the formal properties of the *arithmetic mean*,

$$\alpha(r, s) = \frac{1}{2}(r + s),$$

that is, $\mathfrak{X}$ and $\mathfrak{A} = \langle \mathbb{R}, \leq, \alpha \rangle$ are isomorphic. By Theorem 2.8, the set $\mathcal{S}_\alpha$ of isomorphisms of $\mathfrak{X}$ onto $\mathfrak{A}$ form an interval scale. The function

$$t \to e^t$$

transforms $\alpha$ into the *geometric mean*, $\gamma(r,s) = \sqrt{rs}$, which too has the formal properties of the arithmetic mean, that is, $\mathfrak{G} = \langle \mathbb{R}^+, \leq, \gamma \rangle$ and $\mathfrak{A}$ are isomorphic. It easily follows that the set $\mathcal{S}_\gamma$ of isomorphisms of $\mathfrak{X}$ onto $\mathfrak{G}$ form a log-interval scale.

Given the ambiguity about which operation should be used to represent the data, a natural line of inquiry is to present observers with equal ratio bisection and equal difference bisection tasks on the same stimuli requiring the use of both means. Then it could been seen if the data from the two tasks could be measured in such a way that, simultaneously, the equal ratio operation is represented by the geometric mean and the equal difference operation by the arithmetic mean. In such a paradigm, the first task collects data by asking the observer for stimuli $x$ and $z$ to produce the stimulus $y$ such that the ratio of subjective intensities of $x$ to $y$ is the same as the ratio of subjective intensities of $y$ to $z$. This yields the binary operation $M_R(x,z) = y$ on the set $X$ of stimuli. The second task collects data by asking the observer for stimuli $u$ and $w$ to produce the stimulus $v$ such that the difference of subjective intensities of $u$ to $v$ is the same as the difference of subjective intensities of $v$ to $w$. This yields the binary operation $M_D(u,w) = v$ on the set $X$ of stimuli. Surprisingly, several empirical studies show $M_R = M_D$. For example, Pfanzagl (1968) writes,

> Other inquiries have shown that the values of the arithmetic scale are linearly related to the logarithms of the geometric scale (Torgerson, [1961]; Ekman, 1962; Ekman and Künnapas, 1962a, [1962]b). The natural explanation of this phenomenon is that in these cases the observers are unable to distinguish between arithmetic and geometric bisection: Regardless whether the observers are asked to bisect a given interval from $a$ to $b$ [$a\phi b$] such that the ratio $a:a\phi b$ equals the ratio $a\phi b:b$ or such that the interval from $a$ to $a|b$ [the midpoint of the interval from $a$ to $b$] equal the interval from $a|b$ to $b$, they always perform the same operation. This is also suggested by experiments of Garner (1954). If this were true, [by a previous theorem] a logarithmic relationship would exist between the arithmetic and geometric scales. Intuitively this is obvious: If both operations are in fact identical and the operation is one time mapped into the arithmetic mean and the other time into the geometric mean, the values of the first scale are related to the logarithms of the values of the second scale. (pg. 127)

Theorem 6.1 below provides a theory for the empirical appearance of $M_R = M_D$. It is based on interval scalability and a weakened form of bisymmetry. It relies on the following lemma.[9]

**Lemma 6.1** *Suppose* $\mathfrak{X} = \langle X, \preceq, \ominus \rangle$, $\mathfrak{N} = \langle \mathbb{R}, \leq, \ominus^\star \rangle$, *and* $\mathcal{S}$ *are such that the following five conditions hold for all* $u$, $x$, $y$, *and* $z$ *in* $X$:

1. Commutativity: $\ominus$ *is a binary operation from* $X$ *onto* $X$ *and* $x \ominus y = y \ominus x$.

2. Continuum: $\langle X, \preceq \rangle$ *is a continuum.*

3. Monotonicity: $x \preceq y$ *if and only if* $x \ominus z \preceq y \ominus z$.

4. Right Autodistributivity: $(x \ominus y) \ominus z = (x \ominus z) \ominus (y \ominus z)$.

5. Interval Scalability: $\mathcal{S}$ *is an interval scale family of isomorphisms from* $\mathfrak{X}$ *onto* $\mathfrak{N}$.

*Then* $\ominus^\star$ *is the arithmetic mean.*
**Proof.** The proof of the lemma is elementary and is given in Section 6.5.

Right Autodistributivity, $(x \ominus y) \ominus z = (x \ominus z) \ominus (y \ominus z)$, is a weakened form of Distributivity, $(x \ominus y) \ominus (z \ominus v) = (x \ominus z) \ominus (y \ominus v)$ (which results from the last equation by taking $v = z$), and is a necessary algebraic condition for the arithmetic and geometric means.

**Theorem 6.1** *The observer is presented two tasks. In the first task, she is asked for stimuli $x$ and $z$ from $X$ to produce the stimulus $y$ such that the ratio of subjective intensities of $x$ to $y$ is the same as the ratio of subjective intensities of $y$ to $z$. This yields the binary operation $M_R(x, z) = y$ on the set $X$ of stimuli. In the second task, she is asked for stimuli $u$ and $w$ from $X$ to produce the stimulus $v$ such that the difference of subjective intensities of $u$ to $v$ is the same as the difference of subjective intensities of $v$ to $w$. This yields the binary operation $M_D(u, w) = v$ on the set $X$ of stimuli. Assume that the operations $M_R$ and $M_D$ satisfy the following hypotheses:*

1. Commutativity: $M_R$ *and* $M_D$ *are binary operations from* $X$ *onto* $X$ *and for all* $x$ *and* $y$ *in* $X$, $M_R(x, y) = M_R(y, x)$ *and* $M_D(x, y) = M_D(y, x)$.

2. Continuum: $\langle X, \preceq \rangle$ *is a continuum.*

---
[9]The lemma also follows from a more general theorem of Luce and Narens (1985). The proof presented here is different from Luce and Narens.

3. **Monotonicity.** For all $x$, $y$, and $z$ in $X$, (i) $x \preceq y$ if and only if $M_R(x,z) \preceq M_R(y,z)$; and (ii) $x \preceq y$ if and only if $M_D(x,z) \preceq M_D(y,z)$.

4. **Right Autodistributivity.** For all $x$, $y$, and $z$ in $X$,
$$M_R[M_R(x,y),z] = M_R[M_R(x,z), M_R(y,z)],$$
and
$$M_D[M_D(x,y),z] = M_D[M_D(x,z), M_D(y,z)].$$

5. **Common Interval Scale Family:** $\mathcal{S}$ *is an interval scale family of isomorphisms from* $\mathfrak{X} = \langle X, \preceq, M_R, M_D \rangle$ *onto* $\mathfrak{N} = \langle \mathbb{R}, \leq, M_R^\star, M_D^\star \rangle$.

Then $M_R = M_D$.

**Proof.** Immediate from Lemma 6.1

Theorem 6.1 provides the following rationale for the experimental results showing the identity of bisection operations resulting from equal ratio and equal difference instructions: Observers are using an analog to a 2-point homogeneous, 2-point unique structure to generate their magnitude productions; or equivalently, measuring the stimuli on an interval scale or a scale representationally equivalent to an interval scale such as a log-interval scale to generate their judgments. However, many psychophysical researchers believe or assume—some explicitly, like Torgerson below, and others implicitly—observers are using a ratio scale, or a scale representationally equivalent to a ratio scale, for making magnitude judgments. The conclusion $M_R = M_D$ is not valid if the hypothesis Common Interval Scale Family of Theorem 6.1 is changed to the Common Ratio Family hypothesis that $\mathcal{S}$ is a ratio scale family of isomorphisms from $\mathfrak{X} = \langle X, \preceq, M_R, M_D \rangle$ onto $\mathfrak{N} = \langle \mathbb{R}^+, \leq, M_R^\star, M_D^\star \rangle$.[10] A technical reason as to why ratio scalability does not work for showing $M_R = M_D$ is that $M_R$ and $M_D$ are *3-ary relations*. In general, symmetry imposes weaker restrictions on 3-ary relations than binary relations, and for the argument of Theorem 6.1 to work, a richer symmetry group than those imposed by ratio scalability is needed. However, a result (Theorem 6.3 below) similar to Theorem 6.1 holds for the magnitude production of ratios and differences with a Common Ratio Scale Family assumption, because ratios and differences are *binary relations*.

---

[10] Such a situation occurs when $M^\star(R)$ is the geometric mean and $M_D^\star$ is the arithmetic mean. Note that in this case $\langle X, \preceq, M_D \rangle$ is not, according to Definition 2.14, a continuous bisection structure, because its isomorphic copy $\langle \mathbb{R}^+, \leq, M_D^\star \rangle$ does not satisfy Solvability (Definition 2.14). For example, there does not exist $v$ in $\mathbb{R}^+$ satisfying,
$$\frac{.25 + .25}{2} = \frac{v+1}{2}.$$

## 6.2 TORGERSON'S CONJECTURE

Torgerson (1961) discusses several experiments involving direct judgments of subjective intensity. He concludes,

> These results are all consistent with the notion that the subject perceives only a single quantitative relation between stimuli. When this relation is interpreted as either a psychological distance or a psychological ratio, it can be shown that the subjective magnitudes obey the properties of the corresponding commutative group—the addition group for the distance interpretation and the multiplication group the ratio interpretation. (pg. 205)

In this book, this conclusion of Torgerson is often called *Torgerson's Conjecture,* instead of "Torgerson's conclusion," because it rests heavily on the direct measurement theory of the 1960s, which from this book's perspective is taken to be an unsound theory. In particular, the data Torgerson considered were not rich enough to establish rigorously that "the subjective magnitudes obey the properties of the corresponding commutative group— the addition group for the distance interpretation and the multiplication group the ratio interpretation." However, theoretical and experimental results presented later in this chapter suggest Torgerson's Conjecture is correct for the kinds of experimental situations cited in his article.

Torgerson's Conjecture has various interpretations in the literature with the most common one being a variant of "judgments of ratios are judgments of differences." However, the Conjecture as formulated above by Torgerson also includes, "it can be shown that the subjective magnitudes obey the properties of the corresponding commutative group—the addition group for the distance interpretation and the multiplication group the ratio interpretation." The mathematical analysis of the Conjecture developed in this section include these group-theoretic properties. For the purposes of this section, the following reformulation of the Conjecture is employed:

**Reformulation of Torgerson's Conjecture.** Let $R(x,y)$ is the magnitude function for the instruction, "The ratio of $y$ to $x$ is $p$," and $D(x,y)$ is the magnitude function for the instruction, "The difference between $y$ and $x$ is $q$." Let (1) and (2) be the following propositions:

(1) Subjective intensity for both ratios and differences are measured on the same ratio scale, $\mathcal{S}$.

(2) $\varphi \in \mathcal{S}$ and $R$ is represented by $\varphi$ as a multiplication by a positive real.

Torgerson conjectured and found empirical support for the following:

**Conjecture**: If (1) and (2) hold, then $D$ is represented by $\varphi$ as a multiplication by a positive real. □

Narens (1996, 1997, 2006) shows the Conjecture is implied by the following proposition: *The judgments of ratios and differences are made on a common ratio scale;* that is, (1) alone implies the Conjecture. This result, which is presented as Statements 1 and 2 of Theorem 6.3, employs the following lemma.

**Lemma 6.2** *Suppose $\langle X, \preceq, P_j \rangle_{j \in J}$ is a continuous structure, $F$ is a strictly $\preceq$-increasing function from $X$ onto $X$, and $\mathcal{S}$ is a ratio scale family of isomorphisms from*

$$\mathfrak{X} = \langle X, \preceq, P_j, F \rangle_{j \in J} \quad \text{onto} \quad \mathfrak{N} = \langle \mathbb{R}^+, \leq, P_j^\star, F^\star \rangle_{j \in J}.$$

*Then there exists $r$ in $\mathbb{R}^+$ such that for all $t$ in $\mathbb{R}^+$,*

$$F^\star(t) = rt.$$

**Proof**. Because $\mathcal{S}$ is a ratio scale, it follows from Lemma 3.1 that the set of symmetries of $\mathfrak{N}$ is the the set of multiplications by positive reals. Thus for all $s$ and $t$ in $\mathbb{R}^+$,

$$F^\star(st) = sF^\star(t). \tag{6.1}$$

By Lemma 1.1, there exists positive reals $r$ and $\gamma$ such that for all $t$ in $\mathbb{R}^+$,

$$F^\star(t) = rt^\gamma. \tag{6.2}$$

Thus by Equations 6.2 and 6.1, for each $s$ in $\mathbb{R}^+$,

$$r(st)^\gamma = F^\star(st) = sF^\star(t) = s(rt^\gamma), \tag{6.3}$$

and therefore $s^\gamma = s$, implying $\gamma = 1$. Thus,

$$F^\star(t) = rt^1 = rt. \quad \square$$

The following theorem provides a structural condition that is used to provide a new empirical test for Torgerson's Conjecture.

**Theorem 6.2** *Suppose*

- $\mathfrak{X} = \langle X, \preceq, X_j, F, H \rangle_{j \in J}$ *is a continuous structure,*
- *$F$ and $H$ are strictly $\preceq$-increasing functions from $X$ onto $X$,*

- $\mathfrak{M} = \langle \mathbb{R}^+, \leq, X_j^\star, F^\star, H^\star \rangle_{j \in J}$ is a numerical structure,
- and $S$ is a ratio scale of isomorphisms of $\mathfrak{X}$ onto $\mathfrak{M}$.

Then $F$ and $H$ commute, that is $F * H = H * F$.

**Proof.** By Lemma 6.2, $F^\star$ and $H^\star$ are multiplications by positive reals and thus $F^\star$ and $H^\star$ commute. Because $\mathfrak{X}$ and $\mathfrak{M}$ are isomorphic, it then follows by Theorem 3.3 that $F$ and $H$ commute. □

**Theorem 6.3 (Torgerson's Conjecture)** *Let $R(x,y)$ is the magnitude function for the instruction, "The ratio of $y$ to $x$ is $p$," and $D(x,y)$ is the magnitude function for the instruction, "The difference between $y$ and $x$ is $q$."* **Suppose:** *$R$ and $D$ are measured on a common ratio scale $S$ and $\varphi \in S$. Then the following three statements are true:*

1. *$R$ is represented by $\varphi$ as a multiplication by a positive real (one of Torgerson's hypotheses).*

2. *$D$ is represented by $\varphi$ as a multiplication by a positive real (Torgerson's Conjecture).*

3. *$R * D = D * R$ (the commutativity of ratios and differences).*

**Proof.** Statements 1 and 2 are immediate consequences of Lemma 6.2. Statement 3 is an immediate consequence of Theorem 6.2.

Statement 3 of Theorem 6.3 is used in the following section to test Torgerson's Conjecture. In view of Theorem 6.3, this can also be seen as a test of the existence of a common ratio scale underlying ratio and difference judgments.

## 6.3 AN EXPERIMENTAL TEST OF TORGERSON'S CONJECTURE

There has been a number of empirical studies in the literature that tested whether the data from judgments of ratios were distinguishable from those of judgments of differences. Most found that they were not. However, the data from many of these experiments were not rich enough to test whether the ratio and difference judgments also had the structural properties characteristic of ratios and the differences.

Many psychophysicists assume that psychological intensities like loudness are measurable by a ratio scale. The following experiment of Ellermeier, Narens, & Dielmann (2003) is based on Theorem 6.2. This theorem shows that if strictly increasing functions $F$ and $H$ on a continuum of stimuli are measured by a common ratio scale of isomorphisms, then they commute. This suggests that a ratio magnitude function $R$ produced through a ratio production instruction should commute with a difference magnitude function $D$ produced through a difference production instruction. Ellermeier et al. (2003) asked observers to make ratio and difference productions. In the ratio condition, observers were presented a tone and asked to make the second tone $p$ times as loud as the first one (denoted by $R_p(t) = x$, where $t$ is the first tone and $x$ is the result of the adjustment) for $p = 2, 3$. Ellermeier et al. (2003) write the following about their choice for the difference instruction:

> we cannot simply say for difference productions "adjust the second tone so that the loudness difference is $p$" without providing a unit. Therefore, the following *difference matching* instruction $D_{a,b}$, $a \prec b$, was implemented where $D_{a,b}(x) = y$ holds if and only if the observer adjusts a stimulus $y$ such that "the difference in loudness between $y$ and $x$ is the same as the difference between $b$ and $a$." (pg. 72)

Two different choices of $a, b$ were employed in the experiment: $50, 58$ dB SPL and $50, 62$ dB SPL. Starting from the base level of 65 dB SPL this yields four tests of the commutativity of $R_p$ and $D_{a,b}$ illustrated in Figure 6.1.

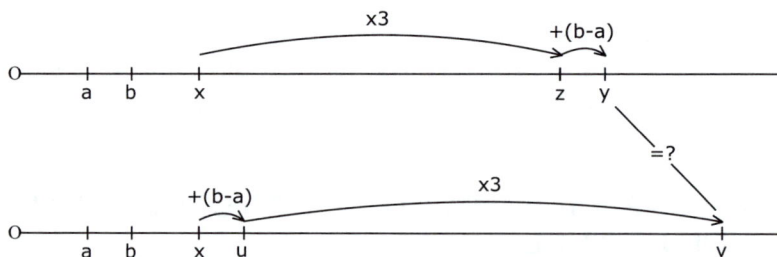

Figure 6.1. Test of the commutativity of $R_3$ and $D_{a,b}$. If commutativity holds, the two orders of chaining these operations should not make a difference, and multiple productions of y and v should be statistically indistinguishable.

Four the the six observers satisfied commutativity between their ratio and difference productions.[11]

---

[11]That two observers were not in line with this outcome is consistent with the observation that some observers sometimes distinguish perceptual ratios and differences (e.g. Schneider, 1980; Birnbaum, 1982).

The commutativity ratios and differences presents a strong experimental constraint, which by Theorem 6.3 provides an empirical test of Togerson's conjecture. The above experiment shows that the constraint holds for most observers. As a test of Torgerson's Conjecture, it has the following distinguishing attributes:

- It uses a new paradigm (commutativity) to demonstrate the Conjecture.

- Unlike Torgerson and many others, it uses indirect rather than direct methods.

- It tests the common underlying ratio scale hypothesis through a qualitative experiment rather than fitting data to a quantitative model.[12]

## 6.4 THEORETICAL CONSIDERATIONS

Many psychophysicists assume that (1) psychological intensities like loudness are measurable by a ratio scale. If, in addition, it is assumed (as Garner, 1954, does) that (2) "There is a true loudness scale, which functions for all kinds of loudness judgments," then it follows by Lemma 6.2 that *any strictly increasing function $F$* corresponding to a loudness judgment from the continuum of stimuli onto itself is represented on the true loudness scale by multiplication by a positive real. Thus assumptions (1) and (2) imply Torgerson's Conjecture. Because of this, they can be taken as a theory for the empirical appearances of the Conjecture in the literature.

Looked at critically, Garner's hypothesis is too strong. Obviously observers can classify sounds into "soft" and "loud," and such a classification implies that the "true loudness scale, which functions for all kinds of loudness judgments," cannot be homogeneous if "soft" and "loud" count as loudness judgments.[13] A more plausible hypothesis is that "there is a true

---

[12]Birnbaum (1978, 1982), Hagerty & Birnbaum (1978), and Veit (1978) have used quantitative fits to the data to argue that the sole underlying operation of magnitude judgment is a subtractive operation.

[13]The following is an argument for this: Suppose $X$ is the set of sounds, $\mathcal{T}$ is a homogeneous scale family used to measure loudness, and $\mathcal{T}$ is family of isomorphisms from a structure $\mathcal{X}$ with domain $X$ onto a numerical structure $\mathfrak{N}$. Suppose $S$ is the nonempty set of soft sounds, $L$ is the nonempty set of loud sounds, and $S \cap L = \varnothing$. Suppose $\mathcal{T}$ appropriately measures $S$ and $L$. Formally this is the following: (i) For all $\varphi$ and $\psi$ in $\mathcal{T}$, (i) $\varphi(S) = \psi(S)$ and $\varphi(L) = \psi(L)$ (that is, $S$ and $L$ are $\mathcal{T}$-representationally meaningful), and (ii) $\psi(S) \cap \psi(L) = \varnothing$ (which follows from the assumption that for all $a \in S$ and all $b \in L$, $\psi(a) < \psi(b)$). Let $a \in S$, $b \in L$ and $\psi \in \mathcal{T}$. By the homogeneity of $\mathcal{S}$, let $\gamma \in \mathcal{S}$ be such that $\gamma(b) = \psi(a)$. Then $\psi(S) \cap \gamma(L) \neq \varnothing$, because $\psi(a)$ is in $\psi(S)$ and $\gamma(b)(= \psi(a))$ is in $\gamma(L)$, and thus $\psi(a)$ is in $\psi(S) \cap \gamma(L)$. However, by (i), $\gamma(L) = \psi(L)$. Thus $\psi(S) \cap \psi(L) \neq \varnothing$, contradicting (ii).

loudness scale that functions for ratio and difference judgments and judgments built out of ratio and difference judgments." Plateau interpreted his midway judgment $M(x,y) = z$ as being built out of the ratio judgments, the ratio of $x$ to $z$ and the ratio of $z$ to $y$.

Suppose $\mathfrak{Y} = \langle Y, \preceq, R_j \rangle_{j \in J}$ is a continuous structure resulting from ratio production or ratio estimation, $k \in J$, and $R_k$ is a $\preceq$-strictly increasing function from $Y$ onto $Y$ such that $y \prec R_k(y)$ for all $y$ in $Y$. Then $\mathfrak{Y}$ cannot be 2-point homogeneous. The reason is that the symmetry group of $\mathfrak{Y}$ is a subgroup of symmetry group of $\mathfrak{Y}_k = \langle Y, \preceq, R_k \rangle$, and it is a consequence of Theorem 4.2 that the symmetry group of $\mathfrak{Y}_k$ is not 2-point homogeneous. Thus, the symmetry group $\mathfrak{Y}$ is not 2-point homogeneous. Therefore $\mathfrak{Y}$ cannot be measured by an interval or log-interval scale family of isomorphisms. Some in the literature have suggested log-interval mathematical models for ratio production and estimation, which runs afoul with the above theoretical argument.

## 6.5 PROOF OF LEMMA 6.1

**Lemma 6.1** *Suppose $\mathfrak{X} = \langle X, \preceq, \ominus \rangle$, $\mathfrak{N} = \langle \mathbb{R}, \leq, \ominus^\star \rangle$, and $\mathcal{S}$ are such that the following five conditions hold for all $u$, $x$, $y$, and $z$ in $X$:*

1. *Commutativity: $\ominus$ is a binary operation from $X$ onto $X$ and $x \ominus y = y \ominus x$.*

2. *Continuum: $\langle X, \preceq \rangle$ is a continuum.*

3. *Monotonicity: $x \preceq y$ if and only if $x \ominus z \preceq y \ominus z$.*

4. *Right Autodistributivity: $(x \ominus y) \ominus z = (x \ominus z) \ominus (y \ominus z)$.*

5. *Interval Scalability: $\mathcal{S}$ is an interval scale family of isomorphisms from $\mathfrak{X}$ onto $\mathfrak{N}$.*

*Then $\ominus^\star$ is the arithmetic mean.*

**Proof.** Because $\mathcal{S}$ is an interval scale, $\ominus^\star$ is invariant under transformations of the form $x \to rx + s$, $r \in \mathbb{R}^+$ and $s \in \mathbb{R}$. By isomorphism, $\ominus^\star$ also satisfies Right Autodistributivity. Therefore, for all $x$ and $y$ in $\mathbb{R}$,

$$(x \ominus^\star y) - y = (x - y) \ominus^\star (y - y) = (x - y) \ominus^\star 0. \qquad (6.4)$$

Let

$$T(x - y) = (x - y) \ominus^\star 0.$$

Then for each $r$ in $\mathbb{R}^+$,

$$rT(x-y) = r[(x-y) \ominus^\star 0] = [r(x-y)] \ominus^\star r0 = [rx - ry] \ominus^\star 0 = T(rx - ry).$$

Because $x$ and $y$ vary over all reals, $z = x - y$ varies over all reals, and thus for all $r$ in $\mathbb{R}^+$ and all $z$ in $\mathbb{R}$,

$$T(rz) = rT(z). \qquad (6.5)$$

Lemma 1.1 applied to Equation 6.5 for positive $z$ then yields: there exists positive reals $t$ and $\gamma$ such that for all $z$ in $\mathbb{R}^+$,

$$T(z) = tz^\gamma. \qquad (6.6)$$

Thus by Equations 6.6 and 6.5, for each $u$ in $\mathbb{R}^+$,

$$t(uz)^\gamma = T(uz) = uT(z) = u(tz^\gamma), \qquad (6.7)$$

and therefore $u^\gamma = u$, implying $\gamma = 1$. Thus for all $0 < z$,

$$T(z) = tz^1 = tz.$$

A similar argument, with the appropriate reformulation of Lemma 1.1 for negative $z$, yields: there exists $s$ in $\mathbb{R}^+$ such that for all $z < 0$,

$$T(z) = sz.$$

Therefore, let $s$ and $t$ be positive reals such that

$$T(z) = \begin{cases} s \cdot z & \text{if } z < 0 \\ 0 & \text{if } z = 0 \\ t \cdot z & \text{if } z > 0. \end{cases}$$

Replacing $z$ by $x - y$ and $T(x - y)$ by $(x - y) \ominus^* 0$ then yields,

$$(x - y) \ominus^* 0 = \begin{cases} s \cdot (x - y) & \text{if } x - y < 0 \\ 0 & \text{if } x - y = 0 \\ t \cdot (x - y) & \text{if } x - y > 0. \end{cases} \qquad (6.8)$$

Because $\ominus^*$ is invariant under the transformation $z \to z + r$,

$$(x - y) \ominus^* 0 = [(x - y) + y] \ominus^* [0 + y] = x \ominus^* y.$$

Therefore, adding $y$ to the left and right sides of Equation 6.8 and using Equation 6.4 then yields,

$$x \ominus^* y = \begin{cases} s \cdot (x - y) + y = s \cdot x + (1 - s) \cdot y & \text{if } x - y < 0 \\ y & \text{if } x - y = 0 \\ t \cdot (x - y) + y = t \cdot x + (1 - t) \cdot y & \text{if } x - y > 0. \end{cases} \qquad (6.9)$$

Note that it follows from Equation 6.9 that $x \ominus^* x = x$ for all $x$ in $\mathbb{R}$. Suppose $u < v$. Then by isomorphism and the monotonicity of $\ominus$, it follows that $u = u \ominus^* u < u \ominus^* v$, and similarly, $u \ominus^* v < v \ominus^* v = v$. Thus

$$\text{if } u < v \text{ then } u < u \ominus^* v < v.$$

Suppose $0 < u$. By isomorphism and the commutativity of $\ominus$, $\ominus^*$ is commutative. The commutativity of $\ominus^*$, Equation 6.9, and $u < 2u$ then yields,

$$s \cdot (u) + (1-s) \cdot 2u = t \cdot (2u) + (1-t) \cdot u,$$

and thus, $s + 2 - 2s = 2t + 1 - t$, which yields, $2 - s = t + 1$, and therefore,

$$s + t = 1.$$

Thus Equation 6.9 can be reformulated as,

$$\text{if } x \leq y \text{ then } x \ominus^* y = s \cdot x + (1-s) \cdot y, \tag{6.10}$$

and

$$\text{if } y \leq x \text{ then } x \ominus^* y = (1-s) \cdot x + s \cdot y. \tag{6.11}$$

In Equations 6.10 and 6.11, $s$ is an element of $\mathbb{R}^+$. It will be shown that $s = \frac{1}{2}$. By Equation 6.10, elements $a$ and $b$ of $\mathbb{R}$ can be found such that

$$a < a \ominus^* b < 0 < b.$$

Then

$$a < a \ominus^* 0 < 0 < b \ominus^* 0 < b.$$

By isomorphism and the right autodistributivity of $\ominus$, it follows that $\ominus^*$ satisfies Right Autodistributivity. Thus

$$(a \ominus^* b) \ominus^* 0 = (a \ominus^* 0) \ominus^* (b \ominus^* 0). \tag{6.12}$$

Applying Equations 6.10 and 6.11 to Equation 6.12 while noting,

$$a < a \ominus^* 0 < 0 < b \ominus^* 0 < b,$$

then yields

$$s \cdot (a \ominus^* b) + (1-s) \cdot 0 = s \cdot (a \ominus^* 0) + (1-s) \cdot (b \ominus^* 0),$$

which yields,

$$s \cdot [s \cdot a + (1-s) \cdot b] = s \cdot [s \cdot a + (1-s) \cdot 0] + (1-s) \cdot [(1-s) \cdot b + s \cdot 0],$$

which in turn yields,
$$s^2 \cdot a + s(1-s) \cdot b = s^2 \cdot a + (1-s)^2 \cdot b,$$
which implies
$$s = 1 - s,$$
and therefore,
$$s = \frac{1}{2}.$$
It then follows from Equations 6.10 and 6.11 that for all $x$ and $y$ in $\mathbb{R}$,
$$x \ominus^\star y = \frac{x+y}{2},$$
showing $\ominus^\star$ is the arithmetic mean. □

# Part II

# Meaningfulness

# Chapter 7

# Meaningfulness Concepts from Measurement Theory

## 7.1 QUANTITATIVE $\mathcal{S}$-MEANINGFULNESS

Stevens (1946) introduced into science the following meaningfulness concept:

**Definition 7.1** Let $\mathcal{S}$ be a scale family on $X$ and $T$ be a $n$-ary relation on $\mathbb{R}$. Then $T$ is said to be *quantitatively $\mathcal{S}$-meaningful* if and only if for all $x_1, \ldots, x_n$ in $X$ and all $\varphi$ and $\psi$ in $\mathcal{S}$,

$$T[\varphi(x_1), \ldots, T(\varphi(x_n))] \text{ iff } T[\psi(x_1), \ldots, T(\psi(x_n))]. \quad \square$$

Stevens believed that the statistics used in the analysis of data should matchup with the methods used to measure the data. He believed that this was accomplished by requiring the results of the statistical analysis to be quantitatively $\mathcal{S}$-meaningful. Other researchers extended the use of quantitative $\mathcal{S}$-meaningfulness to non-statistical contexts. The following examples from Narens (2002a) illustrates two nonstatistical uses of quantitative $\mathcal{S}$-meaningfulness.

**Example 7.1 (Perceived Risk)** Pollatsek and Tversky (1970) proposed a mathematical model of perceived risk based on a clever deduction from simple qualitative assumptions about how perceived risks might combine. For the purposes of this example, only their mathematical model applied to the special case of "simple gambles" is needed.

The notation $(a, p, b)$, where $a$, $b$, and $p$ are real numbers and $0 < p < 1$ stands for the *simple gamble* of receiving $a$ dollars with probability $p$, and $b$ dollars with probability $(1 - p)$. $R(g)$, for a simple gamble $g$, stands for

the participant's perceived risk of of $g$. Pollatsek and Tversky's model says that
$$R(g) = tV(g) - (1-t)E(g),  \qquad (7.1)$$
where $E(g)$ is the expectation of $g$, that is,
$$E(g) = a \cdot p + b \cdot (1-p),$$
and $V(g)$ is the variance of $g$, that is,
$$V(g) = (a - E(g))^2 + (b - E(g))^2,$$
and $0 < t < 1$ is an individual parameter that varies from participant to participant. The intuitive appeal of the model is that,

($i$) the perceived risk of simple gambles increases with variance and declines with positive expectation,

($ii$) individuals vary in their risk perceptions according to how they trade-off increased variance with increase expectation, and

($iii$) the individuals' variability is captured by the single parameter $t$.

Because of the relationship of ($iii$) to ($i$) and ($ii$), the size of $t$ apparently has a psychological interpretation. It will be argued that this is not the case.

For the purpose of argument, suppose that Equation 7.1 holds empirically in an experiment where participants order pairs of simple gambles according to their perceived risks. That is, participants are asked to judge "Is there more risk in gamble $g$ than gamble $h$?," and that standard methods of data analysis show that participants judged that there was more risk in $g$ than $h$ if and only if $R(g) > R(h)$, that is, the experiment confirmed the validity of Equation 7.1. The following is the issue to be investigated: *Given these circumstances, does the model given in Equation 7.1 really measure perceived risk?*

Roskam (1989) gives the following "thought experiment" to show that it does not. Re-label the simple gamble $(a, p, b)$ as $(\$a, p, \$b)$ in order to make explicit that the outcomes are paid in dollars. (Pollatsek and Tversky use simple gambles with dollar-dominated outcomes to illustrate their theory.) The notation $(fc, q, fd)$ will stand for the gamble of receiving $c$ guilders with probability $q$ and $d$ guilders with probability $(1-q)$. (The guilder was the currency of the Netherlands before the emergence of the Euro.) Assume the participant's behavior matches the model; that is, when presented $g = (\$a, p, \$b)$ and $h = (\$c, q, \$d)$, she judges $g$ as more risky than $h$ if and only if $R(g) > R(h)$, where $R(g)$ and $R(h)$ are as defined earlier. Also assume the participant can convert from dollars into guilders without difficulty, and

*Meaningfulness Concepts from Measurement Theory* 91

| | Gamble | $E(g)$ | $V(g)$ | $t$ | $R(g)$ |
|---|---|---|---|---|---|
| $g_1$ | ($1, .5, -$.50) | +0.25 | 0.5625 | .7 | 0.31875 |
| $g_2$ | (f 2.56, .5, -f 1.28) | +0.64 | 3.6864 | .7 | 2.3885 |
| $g_3$ | ($0, .5, -$1.25) | -0.625 | 0.3906 | .7 | 0.4609 |
| $g_4$ | (f 0.0, .5, -f 3.20) | -1.6 | 2.56 | .7 | 2.272 |

Table 7.1: Roskam's Gambles

that she makes her risk judgments based on the underlying probabilities and *her value for the outcomes*. Then her perceived risk of a gamble $g$ in dollars is the same as that of $g'$, where $g'$ is the translation of $g$ into guilders. Such "translation invariance" is a reasonable and necessary condition for perceived risk concepts of those individuals who can translate from dollars into guilders without difficulty. In other words, if "translation invariance" fails in some such individuals, then it is highly questionable that *perceived risk* is being dealt with.

Assume the participant's value of $t$ is .7, and the exchange rate is f 2.56 to $1. Consider the gambles given in Table 7.1. Note that at the exchange rate of f 2.56 to $1, $g_2$ is a translation of $g_1$, and $g_4$ is a translation of $g_3$, and that the values $R(g_2)$ and $R(g_4)$ are computed by applying Equation 7.1 directly to the guilder amounts. Also note that for the guilder gambles $g_2$ and $g_4$, the table uses the same value of $t$ as for dollar gambles.

Table 7.1 displays an inconsistency: Because $R(g_1) < R(g_3)$, the participant should perceive $g_1$ as less risky than $g_3$. Therefore, because $g_2$ is a translation of $g_1$ and $g_4$ is a translation of $g_3$, it follows by the assumption of translation invariance that

the participant should perceive $g_2$ as less risky as $g_4$.

However, in Table 7.1, $R(g_2) > R(g_4)$, which by the model yields

the participant should perceive $g_2$ as more risky than $g_4$.

There is clearly an inconsistency here.

In relating the real-world lottery gambles with outcomes in dollars to the mathematical theory of probability, Pollatsek and Tversky implicitly used the measuring function $\varphi$ that transforms dollar amounts into numbers by $\varphi(\$r) = r$. They did not consider properties of the scale $\mathcal{S}$ that should be used for measuring dollar-dominated lottery gambles, except implicitly that $\varphi \in \mathcal{S}$. The assumption of "translation invariance" for expressing dollar-dominated gambles in terms of guilders implies $2.56\varphi \in \mathcal{S}$. Roskam's gambles show that $\varphi \in \mathcal{S}$ and $2.56\varphi \in \mathcal{S}$ implies $R$ is not quantitatively $\mathcal{S}$-meaningful. □

**Example 7.2 (Averaging of Rating Data)** The data from judges or raters are often used in the sciences, athletics, and other organized social activities as the basis for ranking objects, performances, or individuals. There are many ways to aggregate ratings. Most of those in practice employ some form of arithmetic averaging of individual ratings. The following illustrates a serious problem inherent in the arithmetic averaging of ratio-scaled ratings.

Suppose:

- $R_1$, $R_2$, and $R_3$ are three people rating two performances, $A$ and $B$.

- For $i = 1, \ldots, 3$, $R_i$ uses the measuring function $\varphi_i$ from a ratio scale family $\mathcal{S}_i$ to rate $A$ and $B$.

- $A$ is judged at least as good as $B$, in symbols, $B \preceq A$ if and only if the arithmetic mean of the ratings from $R_1$, $R_2$, and $R_3$ for $A \geq$ the arithmetic mean of the ratings for $B$; otherwise $B$ is judged better than $A$, in symbols, $A \prec B$.

Suppose further that the following data have been collected:

$$\varphi_1(A) = 5 \qquad \varphi_2(A) = 5 \qquad \varphi_3(A) = 5$$
$$\varphi_1(B) = 8 \qquad \varphi_2(B) = 9 \qquad \varphi_3(B) = 1.$$

The arithmetic means of $A$'s and $B$'s ratings are respectively 5 and 6. Thus by the ranking rule just discussed, $A \prec B$. However, to justify this conclusion—and the rule from which it is derived—more needs to be said about the selection of the $\varphi_i$ from $\mathcal{S}_i$. This is where the theoretical problems with arithmetic averaging enter.

In the social and behavioral sciences it is often assumed that for $i \neq j$, the selection of $\varphi_i$ from $\mathcal{S}_i$ is independent from the selection of $\varphi_j$ from $\mathcal{S}_j$. Thus in particular, in the previous case it would then follow that $R_1$ and $R_2$ could have used the same the same measuring functions, $\varphi_1$ and $\varphi_2$, in giving their ratings, while $R_3$ could have used the measuring function $2\varphi_3$ (instead of $\varphi_3$) in giving his. Then the arithmetic mean ranking rule with these new measuring functions would yield $B \prec A$, contradicting the previous overall rankings by the rule using the old measuring functions. This sort of inconsistency makes the following conclusion unavoidable: *If the arithmetic mean ranking rule is to be valid in the situation just discussed with the $\mathcal{S}_i$ being ratio scales, then the selection of measuring functions from the $\mathcal{S}_i$ must be coordinated.*

The coordination of rating measuring functions is a variation of an important problem studied in economics called "the interpersonal comparison of utility." Many prominent economic theorists have concluded that value scales between individuals cannot be validly compared. This idea, when applied to rating situations, says that a systematic coordination of rating measuring functions (e.g., having all raters choose their measuring functions from their ratio scale families so that each rater gives the same rating to a particular object) does not produce a valid comparison of values underlying those ratings. This issue is discussed in some detail in Narens and Luce (1983).

The "problem of interpersonal comparisons" is not that individual ratings for some activity can not be coordinated—for they obviously can. The problem arises from the use of such coordinations as valid methods for intercomparing values across individuals.

Without validly being able to intercompare values, the arithmetic mean loses most of its intuitive appeal as a plausible statistic for determining overall rankings. This suggests looking at other rules for producing overall rankings for cases where the raters' measuring functions are uncoordinated.

Roberts (1985) and others have suggested the geometric mean as a method of producing an ordering of objects in terms of their ratings from judges using measuring functions from ratio scales. In this case $D$ is at least as good as $C$ if and only if the geometric mean of the ratings $d_i$ for $D$ is $\geq$ the geometric mean of the ratings $c_i$ for $C$. Here it is assumed that there are $n$ raters, $i = 1, \ldots, n$, and $c_i$ and $d_i$ come from a measuring function from rater $i$'s ratio scale. Because for positive real numbers $c_1, \ldots, c_n$, $d_1, \ldots, d_n$, and $r_1, \ldots, r_n$,

$$(c_1 \cdot c_2 \cdots c_n)^{\frac{1}{n}} \geq (d_1 \cdot d_2 \cdots d_n)^{\frac{1}{n}}$$
iff
$$(r_1 c_1 \cdot r_2 c_2 \cdots r_n c_n)^{\frac{1}{n}} \geq (r_1 d_1 \cdot r_2 d_2 \cdots r_n d_n)^{\frac{1}{n}},$$

it is easy to see that the geometric mean rule is *meaningful* in the sense that it produces the same ordering between $C$ and $D$ no matter which measuring functions are chosen from $i$'s ratio scale, $i = 1, \ldots, n$. □

For a systematic study of the meaningfulness of averaging and other merging rules, see Aczél & Roberts (1989).

In Part I of this book, the representational theory of measurement was formulated in terms of scale families of isomorphisms onto a numerical structure and representational meaningfulness. In terms of this form of the representational theory, quantitative $\mathcal{S}$-meaningfulness is closely related to representational meaningfulness.

**Theorem 7.1** *Suppose* $\mathfrak{X} = \langle X, Q_j \rangle_{j \in J}$ *is a qualitative structure and* $\mathcal{S}$ *is a family of isomorphisms onto the numerical structure* $\mathfrak{N} = \langle N, T_j \rangle_{j \in J}$. *Let*

$R$ be a $n$-ary relation on the domain of $\mathfrak{X}$ and $T$ be a $m$-ary relation on the domain of $\mathfrak{N}$. Then the following two statements hold:

1. $R$ is $\mathcal{S}$-representationally meaningful (Definition 2.22) if and only if $\varphi(R)$ is quantitatively $\mathcal{S}$-meaningful (Definition 7.1) for each $\varphi$ in $\mathcal{S}$.

2. $T$ is quantitatively $\mathcal{S}$-meaningful if and only if $\varphi^{-1}(T)$ is $\mathcal{S}$-representationally meaningful for each $\varphi$ in $\mathcal{S}$.

**Proof.** (*i*) Suppose $R$ is $\mathcal{S}$-representationally meaningful and $\varphi$ is an arbitrary element of $\mathcal{S}$. Then $\mathcal{S}$ is the set of isomorphisms of

$$\langle X, Q_j, R \rangle_{j \in J} \quad \text{onto} \quad \langle N, T_j, \varphi(R) \rangle_{j \in J} \,.$$

Thus for all $\psi$ and $\theta$ in $\mathcal{S}$ and all $x_1, \ldots, x_n$ in $X$,

$$\varphi(R)[\psi(x_1), \ldots, \psi(x_n)] \quad \text{iff} \quad R[x_1, \ldots, x_n] \quad \text{iff} \quad \varphi(R)[\theta(x_1), \ldots, \theta(x_n)] \,,$$

and thus $\varphi(R)$ is quantitatively $\mathcal{S}$-meaningful.

(*ii*) Suppose $\varphi$ is an arbitrary element of $\mathcal{S}$ and $\varphi(R)$ is quantitatively $\mathcal{S}$-meaningful. Then for for all $\psi$ and $\theta$ in $\mathcal{S}$ and all $x_1, \ldots, x_n$ in $X$,

$$\varphi(R)[\psi(x_1), \ldots, \psi(x_n)] \quad \text{iff} \quad R[x_1, \ldots, x_n] \quad \text{iff} \quad \varphi(R)[\theta(x_1), \ldots, \theta(x_n)] \,,$$

and thus $R$ is $\mathcal{S}$-representationally meaningful.

(*i*) and (*ii*) show Statement 1. Statement 2 follows by letting $R = \varphi^{-1}(T)$ and using Statement 1. $\square$

Special instances of the representational theory of measurement have been used since Helmholtz (1887). Scott and Suppes (1958) formulated the modern version of it. The approach to representational theory used in Part 1 of this book is a variant of Scott and Suppes' version. The difference is that the variant bases measurement on isomorphisms, whereas Scott and Suppes based measurement on a more general kind of structure preserving function. Scott and Suppes commented (1958),

> A primary aim of measurement is to provide a means of convenient computation. Practical control or prediction of empirical phenomena requires that unified, widely applicable methods of analyzing the important relationships between the phenomena be developed. Imbedding the discovered relations in various numerical relational systems is the most important such unifying method that has yet been found. (pp. 116–117)

The kind of imbeddings they employ are homomorphisms. Unlike isomorphisms, homomorphisms need be neither one-to-one or onto functions:

## Meaningfulness Concepts from Measurement Theory

**Definition 7.2 (homomorphism)** Let $\mathfrak{X} = \langle X, Q_j \rangle_{j \in J}$ be a qualitative structure, $\mathfrak{N} = \langle N, T_j \rangle_{j \in J}$, and $Q_j$ and $T_j$ be $n_j$-ary relations, $n_j \geq 0$, on respectively $X$ and $N$. Then $\varphi$ is said to be a *homomorphism from $\mathfrak{X}$ into $\mathfrak{N}$* if and only if $\varphi$ is a function from $X$ into $N$ and for each $j$ in $J$,

(i) if $n_j = 0$, then $\varphi(Q_j) = \varphi(T_j)$; and

(ii) if $n_j > 0$, then for all $x_1, \ldots, x_{n_j}$ in $X$,

$$Q_j[x_1, \ldots, x_{n_j}] \text{ iff } T[\varphi(x_1), \ldots, \varphi(x_{n_j})].\quad \square$$

The Scott-Suppes theory has no meaningfulness concept. Suppes and Zinnes (1963) extended the theory to include quantitative $\mathcal{S}$-meaningfulness. Roberts (1985) and others have applied the Suppes and Zinnes extended form of the representational theory to cover many different kinds of meaningfulness issues in the behavioral sciences.

## 7.2 QUALITATIVE $\mathcal{S}$-MEANINGFULNESS

Pfanzagl (1968) developed a qualitative meaningfulness concept for the homomorphism version of the representational theory that is similar in many respects to quantitative $\mathcal{S}$-meaningfulness. The following definition extends Pfanzagl's development to nonrepresentational theories of measurement, including Stevens' theory (Stevens, 1946, 1951).

**Definition 7.3 (qualitative $\mathcal{S}$-meaningfulness)** Let $\mathcal{S}$ be a scale family on $X$ and $R$ be a $n$-ary relation on $X$. Then $R$ is said to be *qualitatively $\mathcal{S}$-meaningful* if and only if there exists a $n$-ary relation $T$, $n \geq 0$, such that for all $\varphi$ in $\mathcal{S}$,

(i) if $n = 0$ (that is, $R \in X$), then $\varphi(R) = T$, and

(ii) if $n > 0$, then for all $x_1, \ldots, x_n$ in $X$,

$$R[x_1, \ldots, x_n] \text{ iff } T[\varphi(x_1), \ldots, \varphi(x_n)].\quad \square$$

In the representational theory, there are many equally valid numerical representing structures for a given qualitative structure. As following example shows, this clearly presents a problem for use of qualitative $\mathcal{S}$-meaningfulness in the homomorphism version of the theory.

**Example 7.3** Let

- $X = \{a, b, c\}$, $\preceq$ be a total ordering on $X$ such that $a \prec b \prec c$ and $\mathfrak{X} = \langle X, \prec \rangle$,

- $\mathfrak{N} = \langle \{1,2,3\}, < \rangle$,
- $\mathfrak{M} = \langle \mathbb{R}^+, < \rangle$,
- $R(z)$ be the following relation on $X$: $R(z)$ if and only if $z = b$,
- and $U(z)$ be the following relation on $\{1,2,3\}$: $U(z)$ if and only if $z = 2$.

Then the set $\mathcal{S}$ of homomorphisms of $\mathfrak{X}$ into $\mathfrak{N}$ consists of the single function $\varphi$, where $\varphi(a) = 1$, $\varphi(b) = 2$, and $\varphi(c) = 3$. Thus $R$ is qualitatively $\mathcal{S}$-meaningful, because for each $x$ in $X$,

$$R(x) \text{ iff } U(\varphi(x)).$$

Let $\mathcal{T}$ be the set of homomorphisms of $\mathfrak{X}$ into $\mathfrak{M}$. It will be shown by contradiction that $R$ is not qualitatively $\mathcal{T}$-meaningful. Suppose $R$ were $\mathcal{T}$-meaningful. Then a relation $T$ on $\mathbb{R}^+$ can be found such that for all $x$ in $X$ and all $\psi$ in $\mathcal{T}$,

$$R(x) \text{ iff } T(\psi(x)). \tag{7.2}$$

Let $\varphi$ be as above. Then $\varphi$ is in $\mathcal{T}$. Thus applying Equation 7.2 to $\varphi$ and $c$ yields,

$$R(c) \text{ iff } T(\varphi(c)) \text{ iff } T(3).$$

By the definition of "$R$", not $R(c)$ holds. Thus

$$\text{not } T(3). \tag{7.3}$$

Let $\theta$ be the following function on $X$:

- $\theta(a) = 1$,
- $\theta(b) = 3$,
- and $\theta(c) = 4$.

Then $\theta$ is in $\mathcal{T}$. Applying Equation 7.2 to $\theta$ and $b$ then yields,

$$R(b) \text{ iff } T(\theta(b)) \text{ iff } T(3).$$

By the definition of "$R$", $R(b)$ holds. Thus

$$T(3),$$

contradicting Equation 7.3. □

# 7.3 ENDOMORPHISM INVARIANCE

To avoid having the meaningfulness of a qualitative relation depend on the choice of the numerical representing structure, various measurement theorists have suggested using endomorphism invariance as a meaningfulness concept.

**Definition 7.4 (endomorphism)** Let $\mathfrak{X} = \langle X, Q_j \rangle_{j \in J}$, where for each $j$ in $J$, $Q_j$ is a $n_j$-ary relation on $X$, $n_j \geq 0$. Then $\alpha$ is said to be an *endomorphism* of $\mathfrak{X}$ if and only if $\alpha$ is a function from $X$ into $X$ such that for all $j$ in $J$,

(i) if $n_j = 0$, then $\alpha(Q_j) = Q_j$, and

(ii) if $n_j > 0$, then for all $x_1, \ldots, x_{n_j}$ in $X$,

$$Q[x_1, \ldots, x_{n_j}] \text{ iff } Q[\alpha(x_1), \ldots, \alpha(x_{n_j})]. \quad \square$$

Note that each symmetry of $\mathfrak{X}$ is an endomorphism of $\mathfrak{X}$, and if an endomorphism of $\mathfrak{X}$ is a one-to-one function that is onto the domain of $\mathfrak{X}$, then it is a symmetry of $\mathfrak{X}$.

**Definition 7.5 (symmetry and endomorphism invariance)** Let $\mathfrak{X} = \langle X, Q_j \rangle_{j \in J}$, where for each $j$ in $J$, $Q_j$ is a $n_j$-ary relation on $X$, $n_j \geq 0$. Let $R$ be a $n$-ary relation on $X$. Then:

- $R$ is said to be *($\mathfrak{X}$-)symmetry invariant* if and only if for each symmetry $\alpha$ of $\mathfrak{X}$ and each $x_1, \ldots, x_n$ in $X$,

$$R[x_1, \ldots, x_n] \text{ iff } R[\alpha(x_1), \ldots, \alpha(x_n)].$$

- $R$ is said to be *($\mathfrak{X}$-)endomorphism invariant* if and only if for each endomorphism $\alpha$ of $\mathfrak{X}$ and each $x_1, \ldots, x_n$ in $X$,

$$R[x_1, \ldots, x_n] \text{ iff } R[\alpha(x_1), \ldots, \alpha(x_n)]. \quad \square$$

**Theorem 7.2** *Let $\mathfrak{X} = \langle X, Q_j \rangle_{j \in J}$, where for each $j$ in $J$, $Q_j$ is a $n_j$-ary relation on $X$, $R$ be a $n$-ary relation on $X$, $n \geq 0$, and $\mathcal{S}$ be a scale of homomorphisms of $\mathfrak{X}$ into the numerical structure $\mathfrak{N}$. Then the following three statements hold:*

1. *If $R$ is qualitatively $\mathcal{S}$-meaningful (Definition 7.3), then $R$ is endomorphism invariant.*

2. *If $R$ is endomorphism invariant, then $R$ is symmetry invariant.*

3. *R is symmetry invariant if and only if it is $\mathcal{S}$-representationally meaningful.* (Note that the definition of "$\mathcal{S}$-representationally meaningful," Definition 2.22, requires each element of $\mathcal{S}$ to be an isomorphism of $\mathfrak{X}$ onto $\mathfrak{N}$.)

**Proof.** Statement 3 follows from Theorem 3.5. The proofs of Statements 1 and Statement 2 are straightforward and are left to the reader. □

Examples can be provided that show that the converses of Statements 1 and 2 of Theorem 7.2 do not hold (e.g., Narens, 1981a, p. 34.)

In terms of this section's terminology, Narens (1981a) makes the following comment about the above qualitative meaningfulness concepts.

> a number of concepts of qualitative meaningfulness have been presented, and the problem remains of deciding which, if any, is the "correct" concept. It is my view that there is no single correct concept of meaningfulness. I believe that in the final analysis the choice of the "correct" meaningfulness concept for a structure will not be determined solely by the structure, but in general will depend upon features of the intended measurement application. What we have today is a handful of successful applications of the various meaningfulness concepts; what is still lacking is a general theory of meaningfulness and inference based upon meaningfulness. The meaningfulness concepts presented above are attempts to abstract the common core of this handful of successful applications, and are not based upon any detailed philosophical analysis, and thus their usefulness and generality are somewhat in doubt. Hopefully in the future someone will find a more direct and comprehensive approach to this important problem.
>
> [For purposes of basing theories of meaningfulness on] the meaningfulness concepts considered, symmetry invariance has the greatest applicability, mainly because the most important structures that appear in measurement have an abundance of symmetries. Endomorphism invariance and qualitative $\mathcal{S}$-meaningfulness, when they do not coincide with symmetry invariance, thus far have had far fewer applications. I also believe that these two latter concepts have inherent difficulties [as bases for theories of meaningfulness], which arise from the fact that representations of the qualitative structure are only required to be *into* (rather than *onto*) the [numerical] representing structure. Interesting enough, it is this "into" property of representations

that make [meaningfulness based on endomorphisms and qualitative $\mathcal{S}$-meaningfulness of qualitative relations] natural concepts for measurement. However, to my knowledge, the practice of using "into" representations for the general measurement case has never been philosophically justified. The situations where "into" representations have been useful are rather special. (pp. 45–47)

# Chapter 8

# Preliminary Set Theory

## 8.1 INTRODUCTION

This chapter presents a few elementary concepts of set theory. These concepts are used in later chapters to formulate new theories of meaningfulness based on a notion of scientific definability.

Throughout this book, sets are treated as entities that exist alongside with other entities that are non-sets. The non-sets may be mathematical entities such as individual real numbers or qualitative entities such as individual space-time points or individual stimuli from a continuum of stimuli.

Set theory has principles that guarantee the existence of sets specified in terms of already existing sets. Some of these involve properties formulated in terms of the formal language $\mathsf{L}(\in, A)$ described next.

## 8.2 THE LANGUAGE $\mathsf{L}(\in, A)$

An important class of propositions about the mathematics of sets are formulable within a simple formal language. This language plays a central role in the developments of concepts in this and later chapters.

**Definition 8.1 (language $\mathsf{L}(\in, A)$)** The *language* $\mathsf{L}(\in, A)$ is the first-order language that has

- the equality symbol $=$ to denote the identity relation,
- the binary relation symbol $\in$ to denote set-theoretic membership,
- and the individual constant symbol $A$ to denote the substantive domain of scientific inquiry, which is taken as a set of "atoms," that is, a set of objects that are not themselves sets.

$\mathsf{L}(\in, \boldsymbol{A})$ is built out of the following set of symbols:

- the individual constant symbol $\boldsymbol{A}$;
- variables: $x_1, \ldots, x_i, \ldots$ ;
- the 2-place nonlogical predicate symbol $\in$;
- logical symbols: the connectives $\neg$ and $\wedge$; the quantifier $\exists$; and the 2-place predicate symbol $=$;
- and the parentheses ( and ).

The *atomic formulas* of $\mathsf{L}(\in, \boldsymbol{A})$ are defined as follows: If $u$ is a variable or individual constant symbol and $v$ is a variable or individual constant symbol, then $(u = v)$ and $(u \in v)$ are atomic formulas of $\mathsf{L}(\in, \boldsymbol{A})$, and all atomic formulas of $\mathsf{L}(\in, \boldsymbol{A})$ have these forms.

The *formulas* of $\mathsf{L}(\in, \boldsymbol{A})$ are defined inductively as follows:

1. Atomic formulas of $\mathsf{L}(\in, \boldsymbol{A})$ are formulas of $\mathsf{L}(\in, \boldsymbol{A})$.

2. If $\Phi$ and $\Psi$ are formulas of $\mathsf{L}(\in, \boldsymbol{A})$, then $\neg \Phi$ and $(\Phi \wedge \Psi)$ are formulas of $\mathsf{L}(\in, \boldsymbol{A})$.

3. If $\Phi$ is a formula of $\mathsf{L}(\in, \boldsymbol{A})$ and $x$ is a variable, then $(\exists x)\Phi$ is a formula of $\mathsf{L}(\in, \boldsymbol{A})$.

4. All formulas of $\mathsf{L}(\in, \boldsymbol{A})$ can be obtained by repeated application of rules 1–3 above.

In interpreting expressions of $\mathsf{L}(\in, \boldsymbol{A})$,

- $\neg$ is to be interpreted as negation or "not,"
- $\wedge$ is to be interpreted as conjunction or "and,"
- $\exists$ is to be interpreted as the existential quantifier "there exists" or "for some,"
- and $=$ is to be interpreted as as the identity relation.

Other logical connectives and quantifiers are defined in terms of these. In particular,

- implication, $\rightarrow$ ("if ... then ... "), is defined by

$$(\Phi \rightarrow \Psi) \text{ iff } \neg(\Phi \wedge \neg \Psi),$$

*Preliminary Set Theory* 103

- logical equivalence, $\leftrightarrow$ ("if and only if"), is defined by

$$(\Phi \leftrightarrow \Psi) \text{ iff } ((\Phi \to \Psi) \wedge (\Psi \to \Phi)),$$

- disjunction, $\vee$ ("or"), is defined by

$$(\Phi \vee \Psi) \text{ iff } \neg(\neg \Phi \wedge \neg \Psi),$$

- and the universal quantifier, $\forall$ ("for all"), is defined by

$$(\forall x)\Phi \text{ iff } \neg(\exists x)\neg\Phi.$$

A precise inductive definition can be given for "the occurrence of a variable $x$ in the formula $\Phi$ is free" and for "the occurrence of a variable $x$ in the formula is bound." It is assumed that the reader has some familiarity with this usage of "free" and "bound." In the formula

$$(((\exists x_1)(x_1 = x_2) \vee (\forall x_1)(\forall x_2)(x_2 \in x_1)) \wedge (x_3 = x_4)),$$

all occurrences of $x_1$ are bound, the first occurrence of $x_2$ (reading left to right) is free whereas the second is bound, and the occurrences of $x_3$ and $x_4$ are free. A variable $x$ in a formula $\Phi$ is said to be *free* (or more precisely, a *free variable* of $\Phi$) if and only if there is an occurrence of the variable $x$ in $\Phi$ that is free.

If $\Phi$ is a formula of $\mathsf{L}(\in, \boldsymbol{A})$ and $y_1, \ldots, y_n$ are variables of $\mathsf{L}(\in, \boldsymbol{A})$, then the notation $\Phi(y_1, \ldots, y_n)$ will indicate that the free variables of $\Phi$ are *among* the variables $y_1, \ldots, y_n$ (in that order). Note that $\Phi(y_1, \ldots, y_n)$ does not say that $y_1, \ldots, y_n$ *are* the free variables of $\Phi$, but merely that each free variable of $\Phi$ occurs in the sequence $y_1, \ldots, y_n$. By this convention, $\Psi(y_1, y_2)$ and $\Psi(y_2, y_1)$ say slightly different things; they both say that the set of free variables occurring in $\Phi$ is a subset $\{y_1, y_2\}$, but they mention these free variables in different orders.

Formulas of $\mathsf{L}(\in, \boldsymbol{A})$ having no free variables are called *sentences*. $\square$

**Convention 8.1** Let $S$ be a nonempty set. By definition, the *restriction of $\in$ to $S$*, $\in\!\upharpoonright\! S$ is the set

$$\{(x, y) \mid x \in y \text{ and } x \in S \text{ and } y \in S\}.$$

By convention, to simplify notation, "$\in\!\upharpoonright\! S$" is often written as "$\in$," for example, in the following "$\langle \mathsf{V}, \in\!\upharpoonright\!\mathsf{V}, A\rangle$" is often written as "$\langle \mathsf{V}, \in, A\rangle$."

Once an interpretation is given to the nonlogical symbols $\in$ and $\boldsymbol{A}$, each sentence of $\mathsf{L}(\in, \boldsymbol{A})$, by the interpretation, is either true or false. More precisely, a *set model* of $\mathsf{L}(\in, \boldsymbol{A})$ is a structure of the form $\langle \mathsf{V}, \in, A \rangle$, where $\mathsf{V}$ is a nonempty set, $\in$ is the relation of set membership restricted to $\mathsf{V}$ (Convention 8.1), and $A$ is a set. The elements of $\mathsf{V}$ are called *entities*. The quantifiers $\exists$ and $\forall$ are assumed to range over the entities, and

- $\exists x$ is interpreted as, "for some entity $x$,"
- and $\forall x$ is interpreted as, "for all entities $x$."

The following interpretations are also made:

- $\in$ is interpreted as $\in$,
- $\boldsymbol{A}$ is interpreted as $A$,
- $=$ is interpreted as logical identity, $=$,
- $\neg$ is interpreted as negation,
- $\wedge$ is interpreted as conjunction,
- $\rightarrow$ is interpreted as implication,
- $\leftrightarrow$ is interpreted as "if and only if,"
- and $\vee$ is interpreted as disjunction.

Under these interpretations, each sentence of $\mathsf{L}(\in, \boldsymbol{A})$ is a statement about $\langle \mathsf{V}, \in, A \rangle$ that is either true or false. However, some formulas of $\mathsf{L}(\in, \boldsymbol{A})$ are not statements about the structure $\langle \mathsf{V}, \in, A \rangle$, for example, the formula $x_1 \in \boldsymbol{A}$. By convention, for each entity $a$ in $\mathsf{V}$, the expression $a \in \boldsymbol{A}$ is considered a statement about $\langle \mathsf{V}, \in, A \rangle$—a true statement if $a \in A$ and a false statement if $a \notin A$. More generally, by convention, if $y_1, \ldots, y_n$ are variables, $\Phi(y_1, \ldots, y_n)$ is a formula of $\mathsf{L}(\in, \boldsymbol{A})$, and $a_1, \ldots, a_n$ are entities of $\mathsf{V}$, then $\Phi(a_1, \ldots, a_n)$ is considered a statement about $\langle \mathsf{V}, \in, A \rangle$.[14] □

**Convention 8.2** This book follows many mathematical practices for simplifying notation. In particular, it is understood that if a formal language, for example, $\mathsf{L}(\in, \boldsymbol{A})$, is used to describe a particular model of it, for example, $\langle \mathsf{V}, \in, A \rangle$, then boldface symbols are often replaced with their interpretations in the model. For example, $x_1 \in \boldsymbol{A}$ is often written as $x_1 \in A$.

---

[14]Our conventions allow for the case where a variable, say $y_1$ does not occur in a formula, say $\Phi(y_2, \ldots, y_n)$, of $\mathsf{L}(\in, \boldsymbol{A})$. Suppose $a_1, \ldots, a_n$ are entities. Then using the conventions just discussed, $\Phi(y_1, \ldots, y_n)$ is also a formula of $\mathsf{L}(\in, \boldsymbol{A})$. To avoid confusion, let $\Phi'(y_1, \ldots, y_n)$ stand for $\Phi(y_1, \ldots, y_n)$. Then, by definition, $\Phi'(a_1, \ldots, a_n)$ is true if and only if $\Phi(a_2, \ldots, a_n)$ is true.

Additional symbols for variables are employed to avoid the proliferation of subscripts. To improve readability, brackets [ and ] are used along with parentheses. Parentheses are often be omitted, for example, $(\forall x)(x = x)$ is often written as $\forall x(x = x)$, and $(\Phi \wedge \Psi)$ as $\Phi \wedge \Psi$. The usual practice of $\rightarrow$ and $\leftrightarrow$ dominating other logical symbols in the interpretation of formulas is observed, for example, $\Phi \wedge \Psi \rightarrow \Theta$ is to be read as $(\Phi \wedge \Psi) \rightarrow \Theta$, and $\Phi \wedge \Psi \leftrightarrow \Psi \wedge \Theta$ as $(\Phi \wedge \Psi) \leftrightarrow (\Psi \wedge \Theta)$. Similarly, connectives dominate quantifiers, for example, $\forall x \Phi(x) \wedge \Psi(x,y)$ is to be read as $[\forall x \Phi(x)] \wedge \Psi(x,y)$. The word "and" is often used in place of $\wedge$, "or" in place of $\vee$, "for all" in place of $\forall$, et cetera. Also the logical symbols of $\mathbf{L}(\in, \boldsymbol{A})$ are often used in expressions outside the language $\mathbf{L}(\in, \boldsymbol{A})$. $\square$

## 8.3 BASIC SET THEORY

The theories of meaningfulness developed in the following chapters are about fragments of science. The domain of a fragment is conceptualized as a set of objects, $A$, and the various mathematical and scientific relationships used in the scientific study of $A$ are conceptualized as belonging to a portion of set theory containing the set $A$. The elements of $A$ are thought of non-sets. Although set theory contains non-sets different from the elements of $A$, the theories of meaningfulness of later chapters are formulated in such a way that the elements of $A$ are the only non-sets that need to be considered in the scientific study of the fragment.

Non-sets are called "atoms." Thus $A$ is a set of atoms. Relations on $A$ of scientific interest and other scientific concepts used in the scientific analysis of the fragment based on $A$ are represented as sets. Different fragments of science are usually based on different sets of atoms.

The developments of this and the following chapters conceptualize a fragment of science as a structure of the form $\langle \mathsf{V}, \in, A \rangle$, where $A$ is a non-empty set of atoms, and $\mathsf{V}$ is a set containing $A$. There are many sets and non-sets that are not in $\mathsf{V}$. In particular, the non-sets of individual integers and individual real numbers are not in $\mathsf{V}$, and the set of integers, the set of real numbers, and the mathematics based on them are also not in $\mathsf{V}$. However, it is argued in the next chapter that $\mathsf{V}$ can be chosen in such a way that it contains isomorphic counterparts for all of the ordinary mathematics used in science. For the purposes of science, these $\mathsf{V}$-counterparts function just as well in the conduction of science as ordinary mathematics.

**Definition 8.2** Elements of the domain of set theory are called *entities*. There are two kinds of entities, *sets* and *non-sets*. The non-sets are called *atoms*. $\square$

**Assumptions about the scientific domain**, $A$. Throughout the remainder of this book, $A$ will denote a set called the *scientific domain*. It is assumed that $A$ satisfies the following three statements.

1. $A \neq \varnothing$.

2. $\varnothing \notin A$.

3. For all entities $a$, if $a \in A$, then for all entities $b$, $b \notin a$. □

Set theory contain principles for guaranteeing the existence of certain kinds of sets in terms of entities known or assumed to exist. The following are three such principles:

1. **Power Set:** For each set $S$, there exists a set $\wp(S)$, called the *power set* of $S$, such that for all entities $T$,

$$T \in \wp(S) \text{ iff } T \subseteq S.$$

2. **Union Set:** For each set $S$, there exists a set $\bigcup S$, called the *union (of the elements of)* $S$, such that for all entities $e$,

$$e \in \bigcup S \text{ iff there exists } T \text{ in } S \text{ such that } e \in T.$$

Sometimes the Union Set is written in other notations, for example,

$$\bigcup_{i=1}^{\infty} B_i = \{x \,|\, x \in B_i \text{ for some positive integer } i\} = \bigcup_{i \in \mathbb{I}^+} B_i.$$

3. **Comprehension Principle:** Let $\varphi(x, x_1, \ldots, x_n)$ be an arbitrary formula of $\mathbf{L}(\in, \mathbf{A})$ with free variables $x, x_1, \ldots, x_n$. Interpret $\in$ as $\in$, $\mathbf{A}$ as the set $A$ of atoms, $\wedge$ as conjunction, and $\neg$ as negation, and let the quantifiers $\forall$ and $\exists$ rang over all entities. Then for each set $S$ there exists a set $T$ such that for all entities $b, b_1, \ldots, b_n$,

$$b \in T \text{ iff } \varphi(b, b_1, \ldots, b_n) \text{ and } b \in S. \qquad (8.1)$$

Thus for each set $S$, an application of the Comprehension Principle produces a subset of $S$. The set $T$ in Equation 8.1 is often written as

$$T = \{x \,|\, \varphi(x, b_1, \ldots, b_n) \text{ and } x \in S\}. \quad □$$

Another important principle of set theory is **Extensionality:** Sets $B$ and $C$ are identical, in symbols, $B = C$, if and only if they have the same elements, that is

$$\forall x (x \in B \text{ iff } x \in C).$$

Preliminary Set Theory

It is easy to verify that the assumptions about the set of atoms and the Principles of Power Set, Union Set, Comprehension, and Extensionality have simple formulations in $\mathbf{L}(\in, \mathbf{A})$. With the help of the following definitions, the concepts of $n$-ary relation, function, domain of a relation, and related concepts also are easily expressed in $\mathbf{L}(\in, \mathbf{A})$.

**Definition 8.3 (ordered pair)** Let $a$ and $b$ be arbitrary entities. Then, by definition,

- $\{a\} = \{x \mid x = a\}$,
- $\{a, b\} = \{x \mid x = a \text{ or } x = b\}$,
- and $(a, b) = \{\{a\}, \{a, b\}\}$.

$\{a, b\}$ is called the *unordered pair* of $a$ and $b$, and $(a, b)$ is called the *ordered pair* of $a$ and $b$. □

**Theorem 8.1** Let $a$, $b$, $c$, and $d$ be entities. If $(a, b) = (c, d)$ then $a = c$ and $b = d$.

**Proof.** Suppose $(a, b) = (c, d)$.

*Case 1.* $a = b$. Then $(a, b) = \{\{a\}\}$, and thus $(a, b)$ has only one element, $\{a\}$. Because $(a, b) = (c, d)$, it follows that $(c, d)$ has $\{a\}$ as its only element. Because by the definition of ordered pair $\{c\} \in (c, d)$, it follows that $\{a\} = \{c\}$. This implies that $a = c$. Because

$$(c, d) = \{\{c\}, \{c, d\}\} = \{\{c\}\},$$

$\{\{c, d\}\} = \{\{c\}\}$, and thus $\{c, d\} = \{c\}$, that is, $c = d$. Thus it has been shown that $a = b = c = d$, and in particular, $a = c$ and $b = d$.

*Case 2.* $a \neq b$. Then $(a, b)$ has more than one element, and therefore $(c, d)$ has more than one element. Therefore $c \neq d$. Because

$$\{\{a\}, \{a, b\}\} = \{\{c\}, \{c, d\}\},$$

it follows that either (*i*) $\{a\} = \{c, d\}$ and $\{a, b\} = \{c\}$, or (*ii*) $\{a\} = \{c\}$ and $\{a, b\} = \{c, d\}$. (*i*) is not possible, because $\{a\}$ is a has only one element and $\{c, d\}$ has more than one element. Therefore (*ii*) holds, and thus $\{a\} = \{c\}$, implying $a = c$. This together with $\{a, b\} = \{c, d\}$ implies $b = d$. □

The following definition extends Definition 8.3 to ordered $n$-tuples.

**Definition 8.4 (ordered $n$-tuple)** For all positive integers $n$ and all entities $b_1, \ldots, b_n$, define the *ordered $n$-tuple* $(b_1, \ldots, b_n)$ as follows:

- if $n = 1$, then $(b_1) = b_1$;
- if $n > 1$, then $(b_1, \ldots, b_n) = ((b_1, \ldots, b_{n-1}), b_n)$. □

The analog to Theorem 8.1,

$$(a_1, \ldots, a_n) = (b_1, \ldots, b_n) \text{ iff } a_i = b_i \text{ for } i = 1, \ldots, n,$$

easily follows from Theorem 8.1 by induction.

**Definition 8.5 ($n$-ary relation)** Let $n$ be a positive integer and $R$ an entity. Then $R$ is said to be a *$n$-ary relation* if and only if $R$ is a nonempty set such that each element of $R$ is an ordered $n$-tuple.

If $R$ is a $n$-ary relation, then $(b_1, \ldots, b_n) \in R$ is often written as $R(b_1, \ldots, b_n)$.

A 2-ary relation is often called a *binary relation*. □

**Definition 8.6** Suppose R is a binary relation. By definition,

(i) $R^{-1} = \{(y, x) \,|\, R(x, y)\}$,

- the *domain* of $R$ is $\{x \,|\, \exists y R(x, y)\}$, and
- the *codomain* of $R$ is $\{y \,|\, \exists x R(x, y)\}$. □

**Definition 8.7 (function)** $F$ is said to be a *function* if and only if $F$ is a binary relation and

$$\forall x \forall y \forall z [(x, y) \in F \wedge (x, z) \in F \rightarrow y = z].$$

Suppose $F$ is a function. Then by definition:

- If $x \in$ domain of $F$, then $F(x)$ is the entity $y$ such that $(x, y) \in F$.
- $F$ is said to be *on* $X$ (or *from* $X$) if and only if $X$ is domain of $F$.
- $F$ is said to be *into* $Y$ if and only if codomain of $F$ is a subset of $Y$.
- $F$ is said to be *onto* $Y$ if and only if the codomain of $F$ is $Y$.
- $F$ is said to be *one-to-one* if and only if for all $x$ and $y$ in the domain of $F$, if $F(x) = F(y)$ then $x = y$. It is easy to show that $F$ is one-to-one if and only if $F^{-1}$ is a function.
- Suppose $F$ is a function and $S$ is a subset of the domain of $F$. Then *the restriction of $F$ to $S$*, in symbols, $F \restriction S$, is the function $g$ with domain $S$ such that for each $x$ in $S$, $f(x) = g(x)$. □

*Preliminary Set Theory* 109

## 8.4 THE SETS V AND P

The set V defined below is a huge set that contains all the mathematical concepts necessary for describing a fragment of science. The set P defined below is a huge subset of V that contains an isomorphic copy of all the mathematics ordinarily used in science.

**Definition 8.8** $(V_n, V_\infty, \mathsf{V}, P_n, P_\infty, \mathsf{P})$ For each non-negative integer $n$, define the sets $V_n$ and $P_n$ through mathematical induction as follows:

- For $n = 0$, $V_0 = A$ and $P_0 = \varnothing$.

- If $V_n$ has been defined, then
$$V_{n+1} = \wp(V_n) \cup V_n \text{ and } P_{n+1} = \wp(P_n) \cup P_n,$$
where $\wp$ is the power set function.

- By definition,
$$V_\infty = \bigcup_{k=0}^{\infty} V_k \text{ and } P_\infty = \bigcup_{k=0}^{\infty} P_k. \quad \square$$

If $A$ is infinite, then $V_\infty$ is a very large and rich set. However, for purposes of the following chapters, a still larger set is needed. This is provided in the following definitions.

For each non-negative integer $n$, define the sets $V_{\infty+n}$ and $P_{\infty+n}$ through mathematical induction as follows:

- For $n = 0$, $V_{\infty+0} = V_\infty$ and $P_{\infty+0} = P_\infty$.

- If $V_{\infty+n}$ has been defined, then
$$V_{\infty+(n+1)} = \wp(V_{\infty+n}) \cup V_{\infty+n} \text{ and } P_{\infty+(n+1)} = \wp(P_{\infty+n}) \cup P_{\infty+n}.$$

- By definition,
$$\mathsf{V} = \bigcup_{k=0}^{\infty} V_{\infty+k} \text{ and } \mathsf{P} = \bigcup_{k=0}^{\infty} P_{\infty+k}. \quad \square$$

**Definition 8.9 (scientific and purely mathematical entities)** V and P play critical roles in the concepts of meaningfulness developed in the following chapters. V is called the *set of scientific entities* and P is called the *set of purely mathematical entities*. $\quad \square$

**Lemma 8.1** $P_\infty \subseteq V_\infty$ and for each non-negative integer $n$, $P_n \subseteq V_n$.

**Proof by mathematical induction.** $P_0 \subseteq V_0$. Suppose $n$ is a non-negative integer and $P_n \subseteq V_n$. Then

$$P_{n+1} = \wp(P_n) \cup P_n \subseteq \wp(V_n) \cup V_n = V_{n+1}.$$

Thus,

$$P_\infty = \bigcup_{n=0}^\infty P_n \subseteq \bigcup_{n=0}^\infty V_n = V_\infty. \quad \square$$

**Lemma 8.2** $P \subseteq V$ and for each non-negative integer $n$, $P_{\infty+n} \subseteq V_{\infty+n}$.

**Proof by mathematical induction.** Almost identical to the proof of Lemma 8.1. $\quad \square$

**Theorem 8.2** Suppose $k$, $m$, and $n$ are arbitrary non-negative integers such that $k \leq m$. Then the following four statements are true:

1. $P \subseteq V$.

2. $P_k \subseteq P_m$, $V_k \subseteq V_m$, $P_{\infty+k} \subseteq P_{\infty+m}$, and $V_{\infty+k} \subseteq V_{\infty+m}$.

3. $P_n \subseteq P_{\infty+m} \subseteq P$ and $V_n \subseteq V_{\infty+m} \subseteq V$.

4. $P_n \subseteq V_n \subseteq V_\infty$ and $P_{\infty+n} \subseteq V_{\infty+n} \subseteq V$.

**Proof.** The proof is direct and is left to the reader. $\quad \square$

**Definition 8.10 (rank of elements of V)** By definition, for each $x$ in V, the *rank of $x$*, in symbols, rank$(x)$, is defined as follows:

- if $x \in V_\infty$, then rank$(x)$ is the least nonnegative integer $k$ such that $x \in V_k$;

- if $x \in V - V_\infty$, then rank$(x)$ is $\infty + k$, where $k$ is the least nonnegative integer such that $x \in V_{\infty+k}$. $\quad \square$

Notice that each element of $V_\infty$ is an element of another element of $V_\infty$, and thus no element of V has rank $\infty + 0$. In particular, $V_\infty$ has rank $\infty + 1$. The following theorem is immediate from prior definitions.

**Theorem 8.3** The following three statements are true:

1. For all $x$ in V, rank$(x) \neq \infty + 0$.

2. For all $x$ in V, rank$(x) = 0$ if and only if $x$ is an atom.

*Preliminary Set Theory*

3. For all $x$ in $\mathsf{V}$, $\mathsf{rank}(x) = n+1$ or $\mathsf{rank}(x) = \infty + n + 1$ *for some nonnegative integer $n$ if and only if $x$ is a set.* □

**Definition 8.11** The ranks of elements of $\mathsf{V}$ are ordered as follows. Let $b$ and $c$ be arbitrary elements in $\mathsf{V}$ and $\mathsf{rank}(b) = \alpha$ and $\mathsf{rank}(c) = \beta$. Then by definition,

- if $b$ and $c$ are in $\mathsf{V}_\infty$, then: $\mathsf{rank}(b) < \mathsf{rank}(c)$ if and only if $\alpha < \beta$;

- if $b$ and $c$ are in $\mathsf{V} - \mathsf{V}_\infty$, then: $\mathsf{rank}(b) < \mathsf{rank}(c)$ if and only if for some integers $k$ and $m$, $0 < k < m$, $\alpha = \infty + k$, and $\beta = \infty + m$;

- if $b$ is in $\mathsf{V}_\infty$ and $c$ is in $\mathsf{V} - \mathsf{V}_\infty$, then $\mathsf{rank}(b) < \mathsf{rank}(c)$. □

**Theorem 8.4** *Suppose $S$ is a set, $S$ is in $\mathsf{V}$, and $\mathsf{rank}(S) = \alpha$. Then*

$$\forall x (x \in S \to \mathsf{rank}(x) < \alpha).$$

**Proof.** By Theorem 8.3, let $\beta$ be such that $\alpha = \beta + 1$, where $\beta$ is either a nonnegative integer or $\beta = \infty + m$ where $m$ is a nonnegative integer. Because $\mathsf{rank}(S) = \beta + 1$, it follows that $S \in V_{\beta+1}$ and $S \notin V_\beta$. Because

$$V_{\beta+1} = \wp(V_\beta) \cup V_\beta,$$

it must be the case that $S \in \wp(V_\beta)$, that is, $S \subseteq V_\beta$. Thus each element of $S$ is in $V_\beta$ and therefore must have rank $\leq \beta$, and thus rank $< \alpha$. □

**Theorem 8.5** *The following four statements are true:*

1. $\forall x (\text{if } x \in \mathsf{P} \text{ then } x \text{ is a set}).$

2. $\forall x \forall y (\text{if } x \in \mathsf{P} \text{ and } y \in x \text{ then } y \in \mathsf{P}).$

3. $\forall x \forall y (\text{if } x \in \mathsf{P} \text{ and } y \in \mathsf{P} \text{ then } (x, y) \in \mathsf{P}).$

4. *Suppose $n$ is a positive integer and $b_1, \ldots, b_n$ in $\mathsf{P}$. Then $(b_1, \ldots, b_n) \in \mathsf{P}$.*

**Proof.** Statement 1 follows immediately from Definition 8.8. Statement 2 follows by an argument similar to the one given in the proof of Theorem 8.4. Statements 3 and 4 immediately follow from the definitions of $\mathsf{P}$, ordered pair, and ordered $n$-tuple. □

## 8.5  FIRST ORDER AND HIGHER-ORDER RELATIONS

Intuitively, the rank of an entity says something about how "abstract" it is. Viewing matters this way, the atoms in $A$ are the least abstract objects, the non-empty subsets of $A$ are at the same level of abstraction as $A$, the power set of $A$ is one level above $A$, the power set of the power set of $A$ is one level above the power set of $A$, et cetera. This way of viewing "abstract" makes $V_\infty$ very abstract, at infinitely many levels of abstraction above the atoms. For convenience, lets call this form of abstraction, *set-theoretic abstraction*.

The rank of a ordered pair of atoms in $A$ is 2. Thus the rank of a binary relation on $A$ is 3. The rank of a 3-ary relation on $A$ is 5. In general, for $n \geq 2$, the rank of a $n$-ary relation on $A$ is $2n - 1$. Thus the set-theoretic level of abstraction of an $n$-ary relation increases with $n$. In particular note that the level the set-theoretic abstraction of a binary relation on $A$ (which is rank 3) is greater than the level of the set-theoretic abstraction of the power set of $A$ (which is rank 2).

**Definition 8.12** Suppose $X$ is a non-empty set. Then $R$ is said to be a *first-order* relation on $X$ if and only if $R \neq \varnothing$ and $R$ is a $n$-ary relation on $X$ for some non-negative integer $n$. Other kinds of relations built up out of first-order relations on $X$ are called *higher-order* relations on $X$. These include 2nd-order relations on $X$, for example, a non-empty $n$-ary relation on a non-empty subset of first-order relations, 3rd-order relations on $X$, for example, a non-empty $n$-ary relation on a non-empty subset of 2nd-order relations on $X$, et cetera.  □

Another way to view "abstraction" is in terms of how abstract relations are. Viewing matters this way, 0-ary relations on $A$, that is, elements of $A$, are less abstract than other first-order relations on $A$. However, any two first-order relations on $A$ that are not elements of $A$ are considered to be at the same level of abstraction. Higher-order relations include first-order relations. Higher-order relations that are not first-order are considered more abstract than first-order relations. Levels of abstraction among higher-order relations can been formulated. However, for the purposes of this book, only the distinction between first- and higher-order is needed.

Note that by the above usage, all first-order relations are higher-order, but some higher-order relations are not first-order.

In the foundations of science, the fragment of science under consideration is often characterized as a structure of the form,

$$\langle X, R_j \rangle_{j \in J},$$

where for each $j \in J$, $R_j$ is a high-order relation on $X$. In most characterizations, the $R_j$ are first-order and observable. Higher-order relations

*Preliminary Set Theory* 113

that are not first-order, if they appear at all, usually appear in some axiom about the fragment of science under consideration. For example, in the characterization of a continuous extensive structure, $\langle X, \preceq, \oplus \rangle$, $\preceq$ and $\oplus$ are first-order relations. However, higher-order relations that are not first-order appear as part of the the axiom of Dedekind Completeness (Definition 2.15).

# Chapter 9

# Scientific Topics

## 9.1 PRINCIPLES FOR SCIENTIFIC TOPICS

Fundamental to the theories presented in this and subsequent chapters is that certain relations and concepts belong to a fragment of science and others do not. Those that belong are called "meaningful" (with respect to the fragment) and those that do not are called "meaningless" (with respect to the fragment). A *theory of meaningfulness* consists of giving necessary conditions for meaningfulness.

Because theories of meaningfulness are metascientific and philosophical, different theories arise naturally out of various epistemological and metaphysical positions typically taken about science. Several formal theories of meaningfulness are presented and discussed in Narens (2002a). The framework and theories presented in this and later chapters are simplified descriptions of a few of those. The theories are based on precise descriptions of "scientific topic" and "meaningfulness." They are designed to satisfy the following five intuitive principles.

**Principle 1.** The domain of the scientific topic is a qualitative set $X$.

**Principle 2.** The scientific topic is determine by a *structure of primitives* $\langle X, Q_j \rangle_{j \in J}$, where each $Q_j$ is a higher-order relation based on $X$.

**Principle 3.** The structure of primitives and each of its primitives belong to the scientific topic.

**Principle 4.** The scientific topic is closed under "scientific definition;" that is, if $b_1, \ldots, b_n$ belong to the topic and $b$ is defined "scientifically" in terms of $b_1, \ldots, b_n$, then $b$ belongs to the scientific topic.

**Principle 5.** A portion of pure mathematics can be used in scientific definitions.

As an example of Principles 1 to 3, consider the geometry of space-time in relativistic physics. This consists of a domain $X$ of space-time points with the relativistic distance function $\rho$ on $X$. $X$ is also the domain for many other qualitative geometries, for example, $X$ with an appropriately defined 4-dimensional euclidean metric, $\delta_4$, on it is a 4-dimensional euclidean geometry, and $X$ with with an appropriately defined 2-dimensional euclidean distance function, $\delta_2$, on it is 2-dimensional euclidean geometry. Each of these geometries is viewed as a different scientific topic. Many 4-dimensional euclidean concepts about relativistic space-time are considered to be outside of space-time relativity theory; they are considered to be "nonrelativistic"—or to to use the language of this chapter—they are are meaningless in the sense that they have no "relativistic interpretation." Many thought experiments in relativistic physics employ in informal ways this use of "meaningless." The meaningful/meaningless distinction also appears as a major ingredient in other methods of inference used in physics, particularly in hypothesis formation and the design of experiments.

Principle 5 reflects a common practice in mathematical science of utilizing a portion of pure mathematics in the formulations of scientific concepts and the derivations of scientific results. Because of the non-empirical, non-qualitative nature of pure mathematics, the combination of Principles 3 to 5 allow relations and concepts to belong to the scientific topic that are neither qualitative nor empirical. Thus a concept of "meaningfulness," based in part on Principles 3 to 5, should not identified with either "empiricalness" or "qualitativeness." In the literature, identifications of meaningfulness concepts with qualitativeness or empiricalness have caused confusion. Section 11.5 discusses more fully the differentiation between meaningfulness and empiricalness.

The portion of mathematics allowable in Principle 5 depends, in part, on the investigator's philosophies of science and mathematics. For example, some may believe that entities of pure mathematics whose existence depend on the Axiom of Choice—a highly nonconstructive and infinitistic principle of set theory—should not be allowed in determining the entities that belong to the topic of science under consideration, whereas the use of real numbers and various operations on them are allowed. Others may believe that all of ordinary mathematics may be used in describing what belongs to the topic. In principle, the portion of mathematics referred to in Principle 5 may range from none to all of ordinary mathematics.

In Section 9.3, Principles 1 to 4 are given interpretations in $\langle \mathsf{V}, \in, A \rangle$. The following section shows that the entities and relations of pure mathe-

Scientific Topics                                                                 117

matics can be identified with elements of P. With this accomplished, Section 9.3 then provides interpretations of Principles 1 to 5 in $\langle \mathsf{V}, \in, A \rangle$.

## 9.2  PURE MATHEMATICS

Let $1_s = \{\varnothing\}$. Then $1_s \in \mathsf{P}$. Suppose $n \in \mathbb{I}^+$ and $n_s$ has been defined and it has been shown that $n_s \in \mathsf{P}$. Let $(n+1)_s = \{n_s\}$. It easily follows from Definition 8.8 that $\{n_s\}$ is in P, and thus that $(n+1)_s$ is in P. Therefore by mathematical induction, $k_s$ is in P for each $k$ in $\mathbb{I}^+$. Let

$$\mathbb{I}_s^+ = \{n_s \mid n \in \mathbb{I}^+\}.$$

It is not difficult to show that

$$\mathbb{I}_s^+ \subseteq \mathsf{P} \text{ and } \mathrm{rank}(\mathbb{I}_s^+) = \infty + 1.$$

Addition $\oplus$, multiplication $\otimes$, and the ordering relation $\leq_s$ are defined on $\mathbb{I}_s^+$ as follows: For all $k$, $m$, and $n$ in $\mathbb{I}^+$,

- $k_s \oplus m_s = n_s$  iff  $k + m = n$,
- $k_s \otimes m_s = n_s$  iff  $k \cdot m = n$,
- $k_s \leq_s m_s$  iff  $k \leq m$.

Because $\mathbb{I}_s^+ \subseteq \mathsf{P}$, it follows from Statement 4 of Theorem 8.5 that $\oplus$, $\otimes$, and $\leq_s$ are in P. Thus

$$\mathfrak{I}_s^+ = (\mathbb{I}_s^+, \leq_s, \oplus, \otimes)$$

is in P. It is not difficult to show that $\mathfrak{I}_s^+$ and $\langle \mathbb{I}^+, \leq, +, \cdot \rangle$ are isomorphic. Thus the structure of ordered positive integers with addition and multiplication has an isomorphic counterpart as an element of P.

A similar result holds for positive rational numbers: Let

$$Z = \{z \mid z = (x, y) \text{ for some } x \text{ and } y \text{ in } \mathbb{I}_s^+\},$$

and let $\sim$ be the binary relation on $Z$ such that for all $x$, $y$, $u$, and $v$ in $\mathbb{I}_s^+$,

$$(x, y) \sim (u, v) \text{ iff } x \otimes v = u \otimes y.$$

It is easy to verify that $Z$ and $\sim$ are in P and $\sim$ is an equivalence relation on $Z$. Let $Z^\sim$ be the set of $\sim$-equivalence classes. For each $\alpha$ in $Z$, let $\alpha^\sim$ be the $\sim$-equivalence of which $\alpha$ is an element. Then it is easy to show that $Z^\sim$ and its elements are in P. Addition $\oplus'$, multiplication $\otimes'$, and the ordering relation $\leq'_s$ are defined on $Z^\sim$ as follows: For all $(x,y)$, $(u,v)$, and $(w,z)$ in $Z$,

- $(x,y)^\sim \oplus' (u,v)^\sim = (z,w)^\sim$ iff $(x \otimes v) \oplus (u \otimes y) = z$ and $y \otimes v = w$,
- $(x,y)^\sim \otimes' (u,v)^\sim = (z,w)^\sim$ iff $x \otimes u = z$ and $y \otimes v = w$,
- and $(x,y)^\sim \leq'_s (u,v)^\sim$ iff $(x \otimes v) \leq_s (u \otimes y)$.

It is not difficult to show that

$$(Z^\sim, \leq'_s, \oplus', \otimes') \text{ and } \langle Ra^+, \leq, +, \cdot \rangle$$

are isomorphic, where $Ra^+$ is the set of positive rational numbers. Thus, the structure of ordered positive rational numbers with addition and multiplication has an isomorphic counterpart as an element of P.

A more complicated argument shows that the structure of ordered positive real numbers with addition and multiplication has an isomorphic counterpart as an element of P. One can extend this result to all real numbers and all ordinary real analysis based on the real numbers. By considering complex numbers as ordered pairs of real numbers, it is not difficult to show that all ordinary complex analysis based on complex numbers have isomorphic counterparts in P. Similar results hold for geometries, topologies, hilbert spaces, et cetera, that are based on the real or complex numbers. Such extensions are possible, because at any given point in establishing isomorphic counterparts as elements of P, there are infinitely many levels of abstraction (ranks) available for defining new relations or structures. This leads to the following observation that is well-accepted by knowledgeable mathematicians.

**Observation:** *All ordinary purely mathematical operations, relationships, structures, et cetera, have isomorphic counterparts as elements of P.*

This is called an "observation" rather than a "theorem," because "ordinary mathematics" has not been precisely specified.

The mathematics contained in $P_\infty$ is very limited, because it is easy to show that each element of $P_\infty$ is a finite set. Thus if the set of atoms, $A$, is infinite, the mathematics in $P_\infty$ is of very limited use in understanding or describing the science based on $A$. However, because $P_\infty$ is an infinite set, P is very rich in infinitary mathematics, and is very useful in understanding and describing science based on infinite domains.

## 9.3 SET-THEORETIC INTERPRETATION

Principles 1 to 5 are interpreted in $\langle V, \in, A \rangle$ as follows: Elements of V are relations that are useful in the scientific analysis of the fragment of science

# Scientific Topics

under consideration. The fragment is about a scientific topic. The scientific topic is specified by an element of V called a "base structure."

**Definition 9.1 (base structure)** Scientific topics are specified by a *base structure*, $\mathfrak{B} = (A, R_j)_{j \in J}$, where the set of atoms $A$ in V is taken to be the domain of the scientific topic, and the relations $R_j$ are in $V_\infty$, and $J$ is a pure set in P. In the formal language of set theory, $(A, R_j)_{j \in J}$ is the ordered pair $(A, H)$, where $H$ is the function on $J$ such that for each $j$ in $J$, $H(j) = R_j$. By definition, *the higher-order primitives of (the base structure)* $\mathfrak{B}$ consists of $A$, and, for $j \in J$, $R_j$. □

The base structure provides an interpretation for Principles 1 and 2. It is used to generate through a definability principle other relations that belong the scientific topic. The generated relations usually include some qualitative entities, some or all of the purely mathematical entities in P, and some entities, like functions from $A$ onto an element of P, that are are neither qualitative nor purely mathematical.

**Definition 9.2 (the symbols Δ and N)** Throughout the remainder of this book, **Δ** stands for a subset of formulas of $\mathsf{L}(\in, A)$, and **N** stands for a subset of P. □

**Definition 9.3 (scientific definability)** Let $(A, R_j)_{j \in J}$ be a base structure and $e$ be an entity. Then $e$ is said to *scientifically defined in terms of the base structure* $\mathfrak{B} = (A, R_j)_{j \in J}$, **Δ**, *and* **N** if and only if

(*i*) $e = A$,

(*ii*) $e = R_j$ for some $j$ in $J$,

(*iii*) $e = \mathfrak{B}$,

or

(*iv*) there exist

- a formula $\Phi(x, x_1, \ldots, x_m, y_1, \ldots, y_n, z)$ in **Δ**,
- higher-order primitives $b_1, \ldots, b_m$ of the base structure $\mathfrak{B} = (A, R_j)_{j \in J}$,
- and pure sets $S_1, \ldots, S_n$ in **N** such that the following is true about $\langle V, \in, A \rangle$:

$$\Phi(e, b_1, \ldots, b_m, S_1, \ldots, S_n, \mathfrak{B})$$
$$\text{and } \forall z [\text{if } \Phi(z, b_1, \ldots, b_m, S_1, \ldots, S_n, \mathfrak{B}) \text{ then } z = e]. \quad \square$$

Scientific definability provides an interpretation of Principles 3 and 4.[15] Principle 5—the use of a portion of pure mathematics in scientific definitions—is also incorporated in the formal concept of scientific definability given in Definition 9.3 by allowing the pure sets in **N** to be part of the definition process.

The choices of **Δ** and **N** for a fragment of science depend on many things, including on the researcher's philosophy of science and the nature of the scientific problem being investigated.

---

[15]Principle 4 follows for certain kinds of **Δ**, for example, **Δ** = the set of all formulas of **L**($\in$, *A*). **Δ** = the set of all formulas of **L**($\in$, *A*) is the only example of **Δ** considered in detail in this book.

# Chapter 10

# Theories of Meaningfulness

## 10.1 AXIOM OF MEASUREMENT

**Definition 10.1 (M, meaningful, and meaningless)** The symbol M stands for a subset of V (Definition 8.8). Elements of M are said to be *meaningful*. Elements of V − M are said to be *meaningless*. M is called *the set of meaningful entities*. □

**Convention 10.1** Occasionally two meaningfulness concepts are needed at the same time. When this occurs, the notations M and M′ are used to describe them. □

A *theory of meaningfulness* provides necessary or necessary and sufficient conditions for M. The conditions are formulated as axioms. Most of the axiomatic theories relate M with $\Delta$ and **N** (Definition 9.3) and therefore are viewed as theories characterizing scientific topics. One theory, which formulates meaningfulness without direct reference to $\Delta$ or **N**, is an axiomatic version of the Erlanger Program. A theorem shows that it too is a characterization of a specific version of "scientific topic." It is argued in chapter 11 that the equivalence of the Erlanger Program with this version of scientific topic not only provides a new way to understand and generalize invariance under a group of transformations, but also provides an epistemological foundation for important applications of invariance in science and mathematics.

Part I of this book viewed measurement as the assignment of real numbers to qualitative entities. Current scientific measurement is more general: it assigns all sorts of mathematical objects to qualitative entities, for example, complex numbers, quaternions, vectors, points in a metric space, nodes on a graph, et cetera. Part II of this book models measurement as the assignment of elements of V to entities in P. This allows for the scientific domain $A$ to be assigned various kinds of mathematical entities through the measurement process.

Although all of ordinary mathematics have counterparts in P, only a small portion of V can be represented through measurement by isomorphisms onto elements of P. This is because if $A$ is infinite, then there are entities in V that are much larger in size (cardinality) than all the entities in P, and thus cannot be measured by one-to-one functions into P. This sort of difficulty disappears if the meaningfulness theories are formulate for certain larger sets than V or for all sets. In any case, P is large enough to measure any given element of $V_\infty$, which is all that is needed for applications in science.

In principle, each relation of scientific interest is in $V_n$ for some positive integer $n$. Thus the assumption that the domain $A$ and the higher-order relations $R_j$ of a base structure $(A, R_j)_{j \in J}$ are in $V_n$ for some $n \in \mathbb{I}^+$ is not very restrictive from the point of view of science. The indexing set $J$, which by assumption is a pure set, is in general not an element of $V_n$, and cannot be an element of $V_n$ if it is infinite. Thus for a base structure $(A, R_j)_{j \in J}$, the higher-order primitives $R_j$, $j \in J$, are in $V_\infty$, but the higher-order structure $(A, R_j)_{j \in J}$, may not be in $V_\infty$.

Thus meaningfulness in Part II of this book is only considered for entities in V. For applications where this is not reasonable, matters can always be reformulated using larger sets in place of V and P.

**Definition 10.2 (Axiom of Measurement)** The *Axiom of Measurement* is said to hold if and only if there exists a one-to-one function $\varphi$ from $A$ onto an element of P. □

The Axiom of Measurement is used to show the existence of an isomorphism of a base structure (Definition 9.1) onto a structure with domain a pure set. Isomorphisms must preserve logical as well as mathematical structure. In terms of Part II of this book's development, this means that isomorphisms must preserve $\in$, because $\in$ plays a crucial role in formulating counterparts of the logic of higher-order relationships within $\langle V, \in, A \rangle$.

In the following, $D$ and $\varphi$ are such that $D$ is a pure set in P and $\varphi$ is a one-to-one function from $A$ onto $D$. A structure $\mathsf{P}_D$ of higher-order relations based on $D$ is then constructed such that $\varphi$ naturally extends to an isomorphism $\bar{\varphi}$ from $\langle V_\infty, \in \restriction V_\infty, A \rangle$ onto $\langle \mathsf{P}_D, \in', D \rangle$, where $\in'$ is the restriction of $\in$ to a subset of $\mathsf{P}_D$, in symbols, $\in' \subseteq \in \restriction \mathsf{P}_D$.

**Definition 10.3 (the sets $D_{[n]}$ and $\mathsf{P}_D$)** Let $D \in \mathsf{P}$. The sets $\mathsf{P}_D$ and $D_{[n]}$ for nonnegative integers $n$ are defined inductively as follows:

- $D_{[0]} = D$.

# Theories of Meaningfulness

- For nonnegative integers $n$, $D_{[n+1]} = \wp(D_{[n]}) \cup D_{[n]}$.

- $\mathsf{P}_D = \bigcup_{i=0}^{\infty} D_{[i]}$. □

Note that the definition of $\mathsf{P}_D$ mimics the definition of $V_\infty$ with $D$ in place of $A$.

**Definition 10.4 (the extension $\bar{\varphi}$ of $\varphi$)** Suppose $D \in \mathsf{P}$ and $\varphi$ is a one-to-one function from $A$ ($= V_0$) onto $D$ ($= D_{[0]}$). Then $\varphi$ is extended to a function $\bar{\varphi}$ on $V_\infty$ as follows:

- $\bar{\varphi}_{[0]} = \varphi$.

- For nonnegative integers $n$, $\bar{\varphi}_{[n+1]}$ is the function from $V_{n+1}$ into $D_{[n+1]}$ such that for all $x$ in $V_{n+1}$,

  (i) if $x$ is in $V_n$, then $\bar{\varphi}_{[n+1]}(x) = \bar{\varphi}_{[n]}(x)$, and

  (ii) if $x$ is in $V_{n+1} - V_n$, then $\bar{\varphi}_{[n+1]}(x) = \{\bar{\varphi}_{[n]}(y) \,|\, y \in x\}$.

- $\bar{\varphi} = \bigcup_{i=0}^{\infty} \bar{\varphi}_{[i]}$. (Note that $\bar{\varphi}$ is a function from $V_\infty$ into $\mathsf{P}_D$.) □

Recall that by Convention 8.1, the relation $\in$ in structure $\langle V_\infty, \in, A \rangle$ is really $\in\!\upharpoonright\! V_\infty$; that is, formally "$\langle V_\infty, \in, A \rangle$" should be written as

$$\text{``}\langle V_\infty, \in\!\upharpoonright\! V_\infty, A \rangle\text{''}.$$

Convention 8.1 is generally employed throughout this chapter, except in a few instances where additional clarity is desirable.

**Definition 10.5** Let $D \in \mathsf{P}$. Then $\psi$ is said to be an $\in$-*isomorphism from* $\langle V_\infty, \in, A \rangle$ *onto* $\langle \mathsf{P}', \in', D \rangle$ if and only if the following five statements hold:

1. $\mathsf{P}' = \mathsf{P}_D$. (Definition 10.3)

2. $\psi$ is a one-to-one function from $V_\infty$ onto $\mathsf{P}_D$.

3. $\psi\!\upharpoonright\! A$ is a one-to-one function from $A$ onto $D$.

4. $\psi$ is an isomorphism from $\langle V_\infty, \in\!\upharpoonright\! V_\infty, A \rangle$ onto $\langle \mathsf{P}', \in', D \rangle$.

5. $\in' \subseteq \in$.

6. For all sets $a$ in $V_\infty$, $\psi(a) = \{\psi(x) \,|\, x \in a\}$. □

Let $\psi$ be an $\in$ isomorphism from $\langle V_\infty, \in \upharpoonright V_\infty, A\rangle$ onto $\langle \mathsf{P}', \in', D\rangle$. Although $\psi(V_\infty) = \mathsf{P}'$, it need not be the case that $\in' = \in \upharpoonright \mathsf{P}'$. There are two reasons for this: (i) There does not exist $a$ and $b$ in $A$ such that $a \in b$, but there may be $p$ and $q$ in $D$ such that $\psi(a) = p$, $\psi(b) = q$, and $p \in q$. Such a situation makes $\in' = \in \upharpoonright \mathsf{P}'$ impossible. (ii) There may be an element $p$ in $\mathsf{P}'$ such that for some element $q$ in $D$, $p \in q$ and $\psi^{-1}(q)$ is an atom. Therefore, $\psi^{-1}(p) \notin \psi^{-1}(q)$. Thus $p = \psi[\psi^{-1}(p)] \not\in' \psi[\psi^{-1}(q)] = q$. Therefore,

$$p \in q \text{ but } p \not\in' q,$$

that is, $\in' \neq \in \upharpoonright \mathsf{P}'$.

The following theorem characterizes the relationship between the extensions $\bar\varphi$ of one-to-one functions $\varphi$ from $A$ onto $\mathsf{P}_D$ and $\in$-isomorphisms.

**Theorem 10.1** *Suppose $D \in \mathsf{P}$ and $\varphi$ is a one-to-one function from $A$ onto $D$. Then there exists $\in' \subseteq \in$ such that $\bar\varphi$ (Definition 10.4) is an $\in$-isomorphism of $\langle V_\infty, \in, A\rangle$ onto $\langle \mathsf{P}_D, \in', D\rangle$.*

**Proof.** Theorem 10.8. □

**Theorem 10.2** *Suppose $D \in \mathsf{P}$, $\psi$ is an $\in$-isomorphism from $\langle V_\infty, \in, A\rangle$ onto $\langle \mathsf{P}_D, \in', D\rangle$ (Definition 10.5), and $\varphi = \psi \upharpoonright A$. Then $\psi = \bar\varphi$.*

**Proof.** Theorem 10.9. □

The following lemma is an example of $\bar\varphi$ preserving the logical structure of being a $n$-ary relation on $A$.

**Lemma 10.1** *Suppose $D \in \mathsf{P}$, $\varphi$ is a one-to-one function from $A$ onto $D$, $n \in \mathbb{I}^+$, and $R$ is a $n$-ary relation $A$. Then $\bar\varphi(R)$ is a $n$-ary relation on $D$ and for all $a_1, \ldots, a_n$ in $A$,*

$$R(a_1, \ldots, a_n) \text{ iff } \bar\varphi(R)(\bar\varphi(a_1), \ldots, \bar\varphi(a_n)).$$

**Proof.** Suppose $n = 1$. Then $R$ is a set and, by Theorem 10.1, for all $a_1$ in $R$,

$$R(a_1) \text{ iff } a_1 \in R \text{ iff } \bar\varphi(a_1) \in \bar\varphi(R) \text{ iff } \bar\varphi(R)[\bar\varphi(a)].$$

Suppose $n = 2$. Let $x$ and $y$ be arbitrary elements of $V_\infty$. Then the ordered pair $(x, y)$ is in $V_\infty$, and by Definition 10.4,

$$\begin{aligned}\bar\varphi[(x,y)] = \bar\varphi[\{\{x\}, \{x,y\}\}] &= \{\bar\varphi(\{x\}), \bar\varphi(\{x,y\})\} \quad (10.1)\\ &= \{\{\bar\varphi(x)\}, \{\bar\varphi(x), \bar\varphi(y)\}\} = (\bar\varphi(x), \bar\varphi(y)).\end{aligned}$$

Suppose $n > 2$. Let $a_1, \ldots, a_n$ be arbitrary elements of $A$. Because

$$(a_1, \ldots, a_n) = ((a_1, a_2, \ldots, a_{n-1}), a_n),$$

*Theories of Meaningfulness*

it follows from Equation 10.1 that
$$\bar{\varphi}[(a_1,\ldots,a_n)] = (\bar{\varphi}[(a_1,a_2,\ldots,a_{n-1})], \bar{\varphi}(a_n)).$$

A simple inductive argument yields,
$$(\bar{\varphi}[(a_1,a_2,\ldots,a_{n-1})], \bar{\varphi}(a_n)) = (\bar{\varphi}(a_1), \bar{\varphi}(a_2), \ldots, \bar{\varphi}(a_{n-1}), \bar{\varphi}(a_n)).$$

Thus,
$$\bar{\varphi}[(a_1,\ldots,a_n)] = (\bar{\varphi}(a_1), \bar{\varphi}(a_2), \ldots, \bar{\varphi}(a_{n-1}), \bar{\varphi}(a_n)).$$

Therefore, by Theorem 10.1,

$$\begin{aligned}
R(a_1,\ldots,a_n) \quad &\text{iff} \quad (a_1,\ldots,a_n) \in R \\
&\text{iff} \quad \bar{\varphi}[(a_1,\ldots,a_n)] \in \bar{\varphi}(R) \\
&\text{iff} \quad (\bar{\varphi}(a_1),\ldots,\bar{\varphi}(a_n)) \in \bar{\varphi}(R) \\
&\text{iff} \quad \bar{\varphi}(R)[\bar{\varphi}(a_1),\ldots,\bar{\varphi}(a_n)]. \quad \square
\end{aligned}$$

The proof of Lemma 10.1 easily extends to all non-empty sets $X$ in $V_\infty$, that is, if $n \in \mathbb{I}^+$ and $T$ is a $n$-ary relation on $X$, then for all $x_1,\ldots,x_n$ in $X$,
$$T(x_1,\ldots,x_n) \quad \text{iff} \quad \bar{\varphi}(T)(\bar{\varphi}(x_1),\ldots\bar{\varphi}(x_n)).$$
$\bar{\varphi}$ also preserves all the other forms of higher structure in $\langle V, \in, A \rangle$, for example if $S$ is a non-empty set of $n$-ary relations on $X$, then $\bar{\varphi}(S)$ is a non-empty set of $n$-ary relations on $\bar{\varphi}(X)$, et cetera. This allows for $\bar{\varphi}$ to be a basis for forming isomorphisms of higher-order structures with domain $A$ onto higher-order structures with domain a pure set in P.

**Definition 10.6** Let $\mathfrak{B} = \langle A, R_j \rangle_{j \in J}$ be a base structure (Definition 9.1). (Recall by Definition 9.1, $R_j \in V_\infty$ for each $j \in J$). Suppose $D \in \mathsf{P}$. Then $\psi$ is said to be a $\in$-*isomorphism from* $\mathfrak{B}$ *onto* $\mathfrak{N} = \langle D, S_j \rangle_{j \in J}$ if and only if

- for some $\in'$, $\psi$ is an $\in$-isomorphism of $\langle V_\infty, \in, A \rangle$ onto $\langle \mathsf{P}_D, \in', D \rangle$ (Definition 10.5),

- and for each $j$ in $J$, $\psi(R_j) = S_j$. $\quad \square$

**Theorem 10.3 (existence of purely mathematical isomorphic counterparts of the base structure)** *Assume the Axiom of Measurement. By the Axiom of Measurement, let $\varphi$ and $D$ be such that $D \in \mathsf{P}$ and $\varphi$ is a one-to-one function from $A$ onto $D$. Let $\mathfrak{B} = \langle A, R_j \rangle_{j \in J}$ be a base structure (Definition 9.1). Then there exists $S_j$ in $\mathsf{P}_D$ such that $\bar{\varphi}$ an $\in$-isomorphism from $\mathfrak{B}$ onto $\mathfrak{N} = \langle D, S_j \rangle_{j \in J}$.*

**Proof.** By Theorem 10.1, let $\in'$ be such that $\bar{\varphi}$ is an $\in$-isomorphism of $\langle V_\infty, \in, A \rangle$ onto $\langle \mathsf{P}_D, \in', D \rangle$. For each $j$ in $J$, let $S_j = \bar{\varphi}(R_j)$. Then by Definition 10.6, $\bar{\varphi}$ is an $\in$-isomorphism of $\mathfrak{B}$ onto $\mathfrak{N} = \langle D, S_j \rangle_{j \in J}$. $\quad \square$

## 10.2 AXIOM SYSTEM $\mathcal{FST}$

**Definition 10.7 (Axiom System $\mathcal{ST}(\mathfrak{B})$)** *Axiom System $\mathcal{ST}(\mathfrak{B})$ (Scientific Topic for $\mathfrak{B}$) consists of the following three axioms:*

1. *The Axiom of Measurement* (Definition 10.2).

2. $\mathfrak{B} = (A, R_j)_{j \in J}$ is a base structure (Definition 9.1).

3. *Axiom of Scientific Definability:* For all $e$ in V, $e \in$ M (Definition 10.1) if and only if $e$ is scientifically defined in terms of $\mathfrak{B}$, $\Delta$, and N (Definition 9.3). □

**Definition 10.8** Suppose Axiom System $\mathcal{ST}(\mathfrak{B})$. Then M is called the *meaningfulness concept* of $\mathcal{ST}(\mathfrak{B})$, and $\mathfrak{B}$ is called a *base structure for* M. □

Axiom System $\mathcal{ST}(\mathfrak{B})$ is a very weak theory of meaningfulness because it does not specify $\Delta$ or N. The following axiom system allows for the full use of the pure mathematics and higher-order relations in V for selecting meaningful entities.

**Definition 10.9 (Axiom System $\mathcal{FST}$)** *Axiom system $\mathcal{FST}$ (Full Scientific Topic [with base structure $\mathfrak{B}$])consists of the following three axioms:*

1. *Axiom System $\mathcal{ST}(\mathfrak{B})$* (Definition 10.7).

2. *Full Definability:* $\Delta$ = the set of formulas of $\mathsf{L}(\in, A)$.

3. *Full Use of Pure Mathematics:* N = P. □

Axiom System $\mathcal{ST}(\mathfrak{B})$ is the weakest theory of meaningfulness based on scientific definability and $\mathcal{FST}$ is the strongest. There are many theories of meaningfulness in between that are attainable by appropriately selecting $\Delta$ and N. Because $\mathcal{ST}(\mathfrak{B})$ does not specify $\Delta$ and N, it can be viewed as providing necessary conditions for meaningfulness based on scientific definability.

There are other processes for producing theories of meaningfulness that do not use $\mathcal{ST}(\mathfrak{B})$ and specifications of $\Delta$ and N. Some these produce the same meaningful entities as Axiom System $\mathcal{ST}(\mathfrak{B})$ with $\Delta$ and N appropriately specified; others do not (Narens, 2002a). One set of processes identifies meaningfulness with invariance concepts. The following section shows that one of these, the Erlanger Program (chapter 3), is equivalent to Axiom System $\mathcal{FST}$.

# Theories of Meaningfulness

## 10.3 MEANINGFULNESS AND THE ERLANGER PROGRAM

### 10.3.1 Invariance under Extensions of Permutations

The Erlanger Program (chapter 3) identified a geometry with its transformation group, and the entities belonging to a geometry with the invariants under its transformation group. A theory of meaningfulness results by taking the invariants as the meaningful entities. In order to investigate this theory thoroughly, the higher-order invariants and their logical structure need to be accounted for. This and the following section provides such an accounting by developing a theory of higher-order invariants in terms of permutations on $A$.

**Definition 10.10** $f$ is said to be a *permutation* on $A$ if and only if $f$ is a one-to-one function from $A$ onto $A$. □

Let $f$ be permutation on $A = V_0$. The following is a natural way of extending $f$ to $V_1 = \wp(V_0) \cup V_0$: For each subset $x$ of $A$, let
$$f(x) = \{f(a) | a \in x\}.$$
Expanding on this idea, permutations on $A$ are extended to $V$ as follows:

**Definition 10.11** ($\bar{f}$) Let $f$ be a permutation on $A$. By mathematical induction, $f$ is extended to a function $\bar{f}$ on $V$ as follows: For each $x$ in $V_0 = A$, let
$$f_0(x) = f(x).$$
For each nonnegative integer $n$ and each $y$ in $V_{n+1}$, if $y \in V_n$ let
$$f_{n+1}(y) = y,$$
and if $y \in (V_{n+1} - V_n)$ let
$$f_{n+1}(y) = \{f_n(z) | z \in y\}.$$
Let
$$f_\infty = \bigcup_{i=0}^{\infty} f_i.$$
Then $f_\infty$ is a function on $V_\infty$ ($= \bigcup_{i=0}^{\infty} V_i$). For each nonnegative integer $n$ and each $y$ in $V_{\infty+(n+1)}$, if $y \in V_{\infty+n}$ let
$$f_{\infty+(n+1)}(y) = f_{\infty+n}(y),$$

and if $y \subset (V_{\infty+(n+1)} \ \ V_{(\infty+n)})$ let

$$f_{\infty+(n+1)}(y) = \{f_{(\infty+n)}(z) \mid z \in y\}.$$

Let

$$\bar{f} = \bigcup_{i=0}^{\infty} f_{\infty+i}.$$

$\bar{f}$ is called the *extension* of $f$ to V. □

Let $f$ be a permutation on $A$. Note that by Definition 10.11,

$$\bar{f}(\varnothing) = \{f(x) \mid x \in \varnothing\} = \varnothing.$$

The following lemma follows immediately from Definition 10.11 by induction.

**Lemma 10.2** *Let $f$ be a permutation on $A$, and let $\bar{f}$ be the extension of $f$ to V. Then for each set $S$ in V, $\bar{f}(S) = \{\bar{f}(x) \mid x \in S\}$.* □

**Theorem 10.4** *Let $f$ be a permutation on $A$, and let $\bar{f}$ be the extension of $f$ to V. Then the following nine statements hold:*

1. $\bar{f}(A) = A$

2. $\bar{f}$ *is a one-to-one function on* V.

3. $\bar{f}^{-1} = \overline{f^{-1}}$.

4. $\bar{f}$ *is onto* V.

5. *For all $x$ and $y$ in V, $x \in y$ iff $\bar{f}(x) \in \bar{f}(y)$.*

6. *For all $x$ in P, $\bar{f}(x) = x$.*

7. *For all ordered $n$-tuples $(b_1, \ldots, b_n)$ of elements of* V,

$$\bar{f}((b_1, \ldots, b_n)) = (\bar{f}(b_1), \ldots, \bar{f}(b_n)).$$

8. *If $f$ is the identity function on $A$, then $\bar{f}$ is the identity function on* V.

9. *If $g$ is a permutation on $A$, then $f * g$ is a permutation on $A$ and*

$$\bar{f} * \bar{g} = \overline{f * g}.$$

**Proof.** Theorem 10.10 □

## 10.3.2 ∈-Symmetries

**Theorem 10.5** *Let $f$ be a permutation on $A$ and $\alpha$ be a symmetry of $\langle V, \in, A\rangle$. Then the following two statements hold:*

1. *$\bar{f}$ is a symmetry of $\langle V, \in, A\rangle$.*

2. *Let $g$ be the function on $A$ such that for all $x$ in $A$, $g(x) = \alpha(x)$. Then $g$ is a permutation of $A$ and $\bar{g} = \alpha$.*

**Proof.** 1. By Statements 2 and 4 of Theorem 10.4, $\bar{f}$ is a one-to-one function from $V$ onto $V$. By Statement 1, $\bar{f}(A) = A$, and by Statement 5 for all $x$ and $y$ in $V$,
$$x \in y \text{ iff } \bar{f}(x) \in \bar{f}(y).$$
Thus $\bar{f}$ is a symmetry of $\langle V, \in, A\rangle$.

2. Because $\alpha$ is a symmetry, it is a one-to-one function. Because $\alpha(A) = A$ and for all $x \in A$,
$$\alpha(x) \in \alpha(A) = A,$$
it follows that $\alpha$ restricted to $A$ is a one-to-one function from $A$ into $A$. Similarly, $\alpha^{-1}$ is a symmetry, and thus $\alpha^{-1}$ restricted to $A$ is a one-to-one function from $A$ into $A$. Therefore, because for all $x$ in $A$, $x = \alpha[\alpha^{-1}(x)]$, $\alpha$ is a one-to-one function from $A$ onto $A$. Thus $g = \alpha \upharpoonright A$ is a permutation on $A$, and $g(x) = \alpha(x)$ for all $x$ in $A$. Therefore $\bar{g}$ and $\alpha$ agree on all elements of $V$ of rank 0. Suppose Statement 2 is false. A contradiction will be shown. Let $y$ be an element of $V$ of smallest rank such that $\bar{g}(y) \neq \alpha(y)$. rank$(y) > 0$, and therefore $y$ is a set. Then, because by Theorem 8.4, the ranks of elements of $y$ are less than the rank of $y$, it follows from the choice of $y$ that for all $z \in y$,
$$\bar{g}(z) = \alpha(z).$$
Therefore, by Lemma 10.2,
$$\bar{g}(y) = \{\bar{g}(z) \mid z \in y\} = \{\alpha(z) \mid z \in y\} = \alpha(y),$$
contradicting the choice of $y$ as an element of $V$ such $\bar{g}(y) \neq \alpha(y)$. □

Because higher-order relations and concepts in $V$ are defined in terms of $\in$, symmetries of higher-order relations and concepts in $V$ must preserve $\in$. The concept of "$\in$-symmetry" is designed to accomplish this.

**Definition 10.12** $\beta$ is said to be an *$\in$-symmetry* if and only if $\beta$ is a symmetry of $\langle V, \in, A\rangle$.

Let $\mathfrak{B} = \langle A, R_j\rangle_{j \in J}$ be a base structure. Then $\alpha$ is said to be an *$\in$-symmetry* of $\mathfrak{B}$ if and only if for some permutation $f$ on $A$,

- $\iota = \bar{f}$,

- and for all $j$ in $J$, $\bar{f}(R_j) = R_j$. □

Note that by Theorem 10.5, each ∈-symmetry of a base structure $\mathfrak{B}$ is an ∈-symmetry.

**Theorem 10.6** *Let* $\mathfrak{B} = \langle A, R_j \rangle_{j \in J}$ *be a base structure and $G$ be the set of ∈-symmetries of* $\mathfrak{B}$. *Then* $\langle G, * \rangle$ *is a group.*

**Proof.** Let $\iota$ be the identity function on $\mathsf{V}$ and $\alpha$ and $\beta$ be arbitrary elements of $G$. By Definition 10.12, let $f$, $g$ and $h$ be permutations on $A$ such that $\alpha = \bar{f}$ and $\beta = \bar{g}$ and $h$ is the identity function on $A$.

Then $h$ is a permutation on $A$, $\bar{h}$ is an ∈-symmetry, and by Statement 8 of Theorem 10.4, $\iota = \bar{h}$, where $\iota$ is the identity function on $\mathsf{V}$. Therefore, $\iota$ is an ∈-symmetry. Because $\iota$ is the identity function on $\mathsf{V}$ and for each $j$ in $J$,
$$\iota(R_j) = R_j,$$
$\iota$ is an ∈-symmetry of $\mathfrak{B}$; that is, $\iota \in G$.

By Statement 9 of Theorem 10.4, $\bar{f} * \bar{g}$ is an ∈-symmetry, and thus $\alpha * \beta$ is an ∈-symmetry. Then $\alpha * \beta$ is in $G$, because for each $j$ in $J$,
$$\alpha * \beta(R_j) = \alpha[\beta(R_j)] = \alpha(R_j) = R_j.$$

By Statement 3 of Theorem 10.4, for each $x$ in $\mathsf{V}$,
$$\alpha[\alpha^{-1}(x)] = x = \alpha[\bar{f}^{-1}(x)] = \alpha[\overline{f^{-1}}(x)],$$
and thus,
$$\alpha^{-1} = \overline{f^{-1}}, \tag{10.2}$$
showing that $\alpha^{-1}$ is an ∈-symmetry. Because for each $j$ in J,
$$\bar{f}(R_j) = R_j = \bar{f} * \bar{f}^{-1}(R_j),$$
it follows by Statement 3 of Theorem 10.4 that
$$R_j = \bar{f}^{-1}(R_j) = \overline{f^{-1}}(R_j) = \alpha^{-1}(R_i),$$
and thus that $\alpha^{-1} \in G$. □

Theorem 10.6 justifies the following definition.

**Definition 10.13** *Let* $\mathfrak{B} = \langle A, R_j \rangle_{j \in J}$ *be a base structure and $G$ be the set of ∈-symmetries of* $\mathfrak{B}$. *Then* $\langle G, * \rangle$ *is called the* symmetry group *of* $\mathfrak{B}$. □

*Theories of Meaningfulness* 131

**Lemma 10.3** *Suppose $D \in \mathsf{P}$ and $\varphi$ and $\psi$ are $\in$-isomorphisms from $\langle V_\infty, \in, A \rangle$ onto $\langle \mathsf{P}_D, \in', D \rangle$ (Definition 10.5). Suppose $\varphi$ and $\psi$ are also $\in$-isomorphisms of the base structure $\mathfrak{B} = \langle A, R_j \rangle_{j \in J}$ onto $\mathfrak{N} = \langle \mathsf{P}_D, S_j \rangle_{j \in J}$. Let $h$ be the restriction of $\varphi^{-1} * \psi$ to $A$. Then $\bar{h}$ is an $\in$-symmetry of $\mathfrak{B}$, and $\bar{h} \upharpoonright V_\infty = \varphi^{-1} * \psi$.*
**Proof.** Lemma 10.5. □

**Lemma 10.4** *Formally the base structure $\mathfrak{B} = \langle A, R_j \rangle_{j \in J}$ is described by $\mathfrak{B} = (A, H)$, where $H$ is the function on $J$ such that for each $j \in J$, $H(j) = R_j$. Let $G$ be the $\in$-symmetry group for $\mathfrak{B}$. Then for each $g$ in $G$,*

$$g(\mathfrak{B}) = g[(A, H)] = (A, H) = \mathfrak{B}.$$

**Proof.** By Statement 7 of Theorem 10.4,

$$g[(A, H)] = (g(A), g(H)) = (A, g(H)).$$

Thus it only needs to be shown that $g(H) = H$. Because

$$H = \{(j, R_j) \mid j \in J\},$$

the following is a consequence of $(i)$ Statement 7 of Theorem 10.4, $(ii)$ $g$ being an $\in$-symmetry of $\mathfrak{B}$ and therefore $g(R_j) = R_j$ for each $j \in J$, and $(iii)$ the elements of $J$ being pure sets and thus by Statement 6 of Theorem 10.4 and Theorem 10.5, $g(j) = j$ for each $j \in J$:

$$\begin{aligned} g(H) &= g(\{(j, R_j) \mid j \in J\}) \\ &= \{(g(j), g(R_j)) \mid j \in J\} = \{(j, R_j) \mid j \in J\} = H. \quad \square \end{aligned}$$

### 10.3.3 Axiom System $\mathcal{E}$

**Definition 10.14 (Axiom System $\mathcal{E}$)** *Axiom System $\mathcal{E}$ (with base structure $\mathfrak{B}$ and meaningfulness concept $\mathsf{M}$) consists of the following three axioms:*

1. *Axiom of Measurement:* There exists a function from $A$ into $\mathsf{P}$.

2. *Base structure Axiom for $\mathcal{E}$:* $\mathfrak{B} = (A, R_j)_{j \in J}$ is a base structure (Definition 9.1).

3. *Axiom $\mathcal{SM}$ (Symmetry Meaningfulness):* For each $e$ in $\mathsf{V}$, $e$ is in $\mathsf{M}$ if and only if $\alpha(e) = e$ for each $\in$-symmetry $\alpha$ of $\mathfrak{B}$. □

**Theorem 10.7 (meaningfulness equivalence of $\mathcal{E}$ and $\mathcal{FST}$)** *Assume*

- Axiom System $\mathcal{E}$ with base structure $\mathfrak{B}$ and meaningfulness concept M′,

- and Axiom System $\mathcal{FST}$ with base structure $\mathfrak{B}$ and meaningfulness concept M.

Then $\mathsf{M} \cap V_\infty = \mathsf{M}' \cap V_\infty$.

**Proof.** Let $\mathfrak{B} = (A, R_j)_{j \in J}$ and $G$ be the $\in$-symmetry group of $\mathfrak{B}$.

Part 1. $\mathsf{M} \subseteq \mathsf{M}'$. Let $e$ be an arbitrary element of M. By Axiom System $\mathcal{FST}$, let

- $\Phi(x, x_1, \ldots, x_n, y_1, \ldots, y_m, z)$ be a formula of $\mathbf{L}(\in, \boldsymbol{A})$,

- $b_1, \ldots, b_n$ be primitives of $\mathfrak{B}$,

- and $S_1, \ldots, S_m$ be elements of P,

such that
$$\Phi(e, b_1, \ldots, b_n, S_1, \ldots, S_m, \mathfrak{B})$$
and for all $c$ in V, if
$$\Phi(c, b_1, \ldots, b_n, S_1, \ldots, S_m, \mathfrak{B}),$$
then $c = e$.

Because $b_1, \ldots, b_n$ are primitives of $\mathfrak{B}$, it follows from Definition 10.12 that for each $\alpha$ in $G$ and each $b_i$, $i = 1, \ldots, n$,
$$\alpha(b_i) = b_i. \tag{10.3}$$

By Lemma 10.4, for each $\alpha$ in $G$,
$$\alpha(\mathfrak{B}) = \mathfrak{B}. \tag{10.4}$$

Because $S_1, \ldots, S_m$ are in P, it follows by Statement 6 of Theorem 10.4 and Theorem 10.5 that for each $\alpha$ in $G$ and each $S_k$, $k = 1, \ldots, m$,
$$\alpha(S_k) = S_k. \tag{10.5}$$

Let
$$\mathfrak{X} = \langle \mathsf{V}, \in, A, b_1, \ldots, b_n, S_1, \ldots, S_m, \mathfrak{B} \rangle.$$

Then by the above and Theorem 10.5, each element of $G$ is a $\in$-symmetry of $\mathfrak{X}$. Although $\mathfrak{X}$ is a structure of sets, it may be viewed formally as a relational structure by viewing $\in$ as a binary relation on V, and $A$, $b_1, \ldots, b_n, S_1, \ldots, S_m, \mathfrak{B}$ as elements of V. Viewed this way, $\mathfrak{X}$ is described by the first-order language,
$$\mathbf{L} = \mathbf{L}(\in, \boldsymbol{A}, \boldsymbol{b_1}, \ldots, \boldsymbol{b_n}, \boldsymbol{S_1}, \ldots, \boldsymbol{S_m}, \mathfrak{B}),$$

*Theories of Meaningfulness* 133

where $\in$ is a binary relation symbol and $A, b_1, \ldots, b_n, S_1, \ldots, S_n, \mathfrak{B}$ are individual constant symbols, and $\mathfrak{X}$ is a model of **L**.

The following is the pivotal idea of the proof. $\mathfrak{X}$ can be viewed two ways: $(i)$ as a structure of sets, or $(ii)$ as a relational structure that is a model of **L**. As a model of **L**, $\in$ is viewed as a binary relation, and $A, b_1, \ldots, b_n, S_1, \ldots, S_m, \mathfrak{B}$ as individual constants. In particular, as a model of **L**, the fact that $A$, $S_k$, $\mathfrak{B}$ and possibly some $b_i$ are sets rather than some other sort of objects is not used. The key point used in the proof is that for each $g$ in $G$, $g$ is both $(i)$ an $\in$-symmetry of $\mathfrak{X}$ when $\mathfrak{X}$ is considered a structure of sets, and $(ii)$ a symmetry of $\mathfrak{X}$ when $\mathfrak{X}$ is considered a model of **L**.

It is well known that symmetries of a relational structure preserve relations defined from the primitives of that structure through an appropriate first-order language (for example, see Corollary 22D on page 93 of Enderton, 1972). Applying this result to

- the language **L**,

- the model $\mathfrak{X}$ of **L**,

- the group $G$ of $\in$-symmetries of *the structure of sets* $\mathfrak{X}$, which = the group of symmetries of *the relational structure* $\mathfrak{X}$, because the latter is a model of **L**,

- the entity $e$ in the domain of $\mathfrak{X}$,

- and to the formula $\Phi(x, b_1, \ldots, b_n, S_1, \ldots, S_m, \mathfrak{B})$ of **L**, which by hypothesis defines $e$ in $\mathfrak{X}$ through the statements about $\mathfrak{X}$,

$$\Phi(e, b_1, \ldots, b_n, S_1, \ldots, S_m, \mathfrak{B})$$

and

$$\forall x[\Phi(x, b_1, \ldots, b_n, S_1, \ldots, S_m, \mathfrak{B}) \to x = e],$$

yields the following: $\beta(e) = e$ for all $\beta$ in $G$. *Viewing $\mathfrak{X}$ as a structure of sets*, it then follows from the axiom of Symmetry Meaningfulness (Definition 10.14) that $e$ is in M$'$. Because $e$ was chosen as an arbitrary of M, this establishes that M $\subseteq$ M$'$.

*Part 2.* M$'$ $\cap$ $V_\infty$ $\subseteq$ M $\cap$ $V_\infty$. By the definitions of $\mathcal{FST}$, the axiom of Scientific Definability, and scientific definability (Definition 9.3),

$$A, R_j \text{ for each } j \in J, \text{ and } \mathfrak{B} \text{ are are in M.} \tag{10.6}$$

By the axiom of Measurement of $\mathcal{FST}$ and Theorem 10.3, let $D \in \mathsf{P}$ and $T_j$, $j \in J$, be such that there exists an $\in$-isomorphism of $\mathfrak{B}$ onto $\mathfrak{N} = (D, T_j)_{j \in J}$. Let

$$S = \{\psi \mid \psi \text{ is an } \in\text{-isomorphism from } \mathfrak{B} \text{ onto } \mathfrak{N}\},$$

and let

$$G = \{\alpha \mid \alpha \text{ is an } \in\text{-symmetry of } \mathfrak{B}\}.$$

It easily follows that

- for each $\alpha$ in $G$ and each $\psi$ in $S$, $\psi * \alpha$ is in $S$,

and from Lemma 10.3 that

- for all $\rho$ and $\varphi$ in $\mathcal{S}$, if $h$ is the restriction of $\rho^{-1} * \varphi$ to $A$, then $\bar{h} \in G$.

Let

$$G' = \{\bar{h} \mid \text{for some } \rho \text{ and } \varphi \text{ in } \mathcal{S}, h \text{ is the restriction of } \rho^{-1} * \varphi \text{ to } A\}. \tag{10.7}$$

Then by the above, $G' \subseteq G$. Because for all $\alpha$ in $G$ and all $\psi$ in $S$, $\psi * \alpha \in S$ and $\psi^{-1}(\psi * \alpha) = \alpha$, it follows that $G \subseteq G'$. Thus

$$G = G'. \tag{10.8}$$

Note that $\mathfrak{B}$ is in M by Equation 10.6, $\mathfrak{N}$ is in M by Full Pure Mathematics, and thus $S$ is in M by Full Use of Pure Mathematics and Full Definability.[16]

Suppose $e$ is an arbitrary element of $\mathsf{M}' \cap V_\infty$. Then $\alpha(e) = e$ for all $\alpha$ in $G$. Let $\varphi$ be in $S$ and $e^\star = \varphi(e)$. Then by the definition of "$\in$-isomorphism" (Definition 10.5), $e^\star$ is in P. Let $\Phi(x)$ be the expression

$$\exists y \, [y \in S \text{ and } y(x) = e^\star].$$

Then $\Phi(e)$, because $\varphi$ is in $S$ and $\varphi(e) = e^\star$. Suppose $e'$ is in V and $\Phi(e')$ is true. Let $\rho$ in $S$ be such that $\rho(e') = e^\star$. Let $h$ be the restriction of $\rho^{-1} * \varphi$ to $A$. Then Equations 10.7 and 10.8, $\bar{h}$ is in $G$. Thus, because by hypothesis $e$ is invariant under $G$, it follows from Lemma 10.3 that

$$e = \bar{h}(e) = \rho^{-1} * \varphi(e) = \rho^{-1}(e^\star) = e',$$

---

[16] Several concepts need to be scientifically defined for a scientific definition of $\mathcal{S}$. For example "$\in$-isomorphism" needs to be given a scientific definition, which in turn requires "$V_\infty$" to be given a scientific definition. None of these are difficult. A formal scientific definition of $\mathcal{S}$ is left to the reader.

# Theories of Meaningfulness

that is,
$$\forall z\, [\Phi(z) \text{ iff } z = e].$$
Therefore, because $S$ is in M, $e$ is in M by Full Definability and Full Use of Pure Mathematics. Because $e$ was chosen as an arbitrary element of $\mathsf{M}' \cap V_\infty$, this establishes that $\mathsf{M}' \cap V_\infty \subseteq \mathsf{M} \cap V_\infty$.

Parts 1 and 2 of this proof show $\mathsf{M}' \cap V_\infty \subseteq \mathsf{M} \cap V_\infty$. □

Narens (2002a) provides a more general version of Theorem 10.7. In the more general version, the meaningfulness concepts based on scientific definability and symmetry invariance apply to all entities rather than just to V, and they are shown to be identical rather than just identical on $V_\infty$. However, Theorem 10.7 is sufficiently strong for the purposes of this book, and it is all that is needed for scientific applications.

Theorem 10.7 shows there are two approaches to the meaningfulness concept of the Erlanger Program: One based on invariance under transformation groups, and the other based on the full use of set-theoretic formulae and the full use of pure mathematics in formulating scientific definitions. The next chapter exploits the fact that these two approaches lead to the same meaningfulness concept for entities of $V_\infty$.

## 10.4 ADDITIONAL PROOFS

**Theorem 10.8 (Theorem 10.1)** *Suppose $D \in P$ and $\varphi$ is a one-to-one function from $A$ onto $D$. Then there exists $\in' \subseteq \in$ such that $\bar{\varphi}$ (Definition 10.4) is an $\in$-isomorphism of $\langle V, \in, A \rangle$ onto $\langle P_D, \in', D \rangle$.*

**Proof.** By induction. *Part 0.* By hypothesis, $\varphi$ is a one-to-one function from $A = V_0$ onto $D = D_{[0]}$. Thus, because $\bar{\varphi}_0 = \varphi$, $\bar{\varphi}_0$ is a one-to-one function from $V_0$ onto $D_{[0]}$. Define $\in'_0$ on $D_{[0]}$ as follows: For all $u$ and $v$ in $D_{[0]}$,
$$u \in'_0 v \text{ iff } u \in v \text{ and } \bar{\varphi}_0^{-1}(u) \in \bar{\varphi}_0^{-1}(v).$$
Because $A = V_0$ consists of atoms, $\in \!\upharpoonright\! V_0 = \varnothing$, and thus $\in'_0 = \varnothing$. The above shows that $\bar{\varphi}_0$ is an isomorphism of $\langle V_0, \in \!\upharpoonright\! V_0, A \rangle$ onto $\langle D_{[0]}, \in'_0, D \rangle$, where $\in'_0 \subseteq \in$.

*Part $n+1$.* Suppose $n$ is a non-negative integer and $\bar{\varphi}_n$ is an isomorphism of $\langle V_n, \in \!\upharpoonright\! V_n, A \rangle$ onto $\langle D_{[n]}, \in'_n, D \rangle$, where $\in'_n \subseteq \in$.

By hypothesis, $\bar{\varphi}_n(A) = D$, and thus, $\bar{\varphi}_{n+1}(A) = D$.

By hypothesis, $\bar{\varphi}_n$ is a one-to-one function from $V_n$ onto $D_{[n]}$. Let $p$ be an arbitrary element of $D_{[n+1]}$. Then $p$ is either in $D_{[n]}$ or $\wp(D_{[n]})$. If it is in $D_{[n]}$, then, by hypothesis, let $a$ in $V_n$ be such that $\bar{\varphi}_n(a) = p$. If $p$ is in $D_{[n+1]} - D_{[n]}$, then $p \subseteq D_{[n]}$, and let
$$a = \{\bar{\varphi}_n^{-1}(y) \,|\, y \in p\}. \tag{10.9}$$

$u$ in Equation 10.9 is well-defined because each $y$ in $p$ is in $D_n$ and by hypothesis $\bar{\varphi}_n$ is a one-to-one function from $V_n$ onto $D_{[n]}$. By Equation 10.9 and Definition 10.4,

$$\bar{\varphi}_n(u) = \{\bar{\varphi}_n[\bar{\varphi}_n^{-1}(y)] \mid y \in p\} = \{y \mid y \in p\} = p.$$

The above shows that $\bar{\varphi}_n$ is onto $D_{[n+1]}$.

To show that $\bar{\varphi}_{n+1}$ is one-to-one, suppose $b$ and $c$ in $V_{n+1}$ are such that $b \neq c$ and $\bar{\varphi}_{n+1}(b) = \bar{\varphi}_{n+1}(c)$. A contradiction will be shown. Because $b \neq c$, either $b - c \neq \varnothing$ or $c - b \neq \varnothing$. Without loss of generality suppose $b - c \neq \varnothing$. Let $e \in (b - c)$. Then by Definition 10.4, $e \in V_n$, $\bar{\varphi}_n(e) \in \bar{\varphi}_{n+1}(b)$, and $\bar{\varphi}_n(e) \notin \bar{\varphi}_{n+1}(c)$. Thus by Definition 10.4, $\bar{\varphi}_{n+1}(b) \neq \bar{\varphi}_{n+1}(c)$, a contradiction.

Because $\bar{\varphi}_{n+1}$ is a one-to-one, let $\in'_{n+1}$ be the binary relation on $D_{[n+1]}$ such that for all $u$ and $v$ in $D_{[n+1]}$,

$$u \in'_{n+1} v \text{ iff } u \in v \text{ and } \bar{\varphi}_{n+1}^{-1}(u) \in \bar{\varphi}_{n+1}^{-1}(v).$$

Then $\in'_{n+1} \subseteq \in$, and thus, because $\bar{\varphi}_{n+1}(A) = D$ and $\bar{\varphi}_{n+1}$ is a one-to-one function from $V_{n+1}$ onto $D_{[n+1]}$, it follows from the definition of "$\in'_{n+1}$" that $\bar{\varphi}_{n+1}$ is an isomorphism of $\langle V_{n+1}, \in \restriction V_{n+1}, A \rangle$ onto $\langle D_{[n+1]}, \in'_{n+1}, D \rangle$.

*Final Part.* It follows from the definition of $\bar{\varphi}$ (Definition 10.4) that

$$\bar{\varphi} = \bigcup_{i=0}^{\infty} \bar{\varphi}_i.$$

Let

$$\in' = \bigcup_{i=0}^{\infty} \in'_i.$$

It then follows that $\bar{\varphi}$ is an isomorphism of $\mathfrak{X} = \langle V_\infty, \in \restriction V_\infty, A \rangle$ onto $\mathfrak{N} = \langle D_{[\infty]}, \in', D \rangle$. Because for each nonnegative integer $i$, $\in'_i \subseteq \in$, it follows that $\in' \subseteq \in$. Thus $\bar{\varphi}$ is an $\in$-isomorphism of $\mathfrak{X}$ onto $\mathfrak{N}$. □

**Theorem 10.9 (Theorem 10.2)** *Suppose $D \in P$ and $\psi$ is an $\in$-isomorphism from $\langle V_\infty, \in, A \rangle$ onto $\langle P_D, \in', D \rangle$ (Definition 10.5), and $\varphi = \psi \restriction A$. Then $\psi = \bar{\varphi}$.*

**Proof.** Suppose $\psi \neq \bar{\varphi}$. A contradiction will be shown. Let $a$ in $V_\infty$ be an element of smallest rank such that $\psi(a) \neq \bar{\varphi}(a)$. $a \notin A$, because by hypothesis $\varphi = \psi \restriction A$. Therefore $a$ is a set. $a \neq \varnothing$, because $\psi(\varnothing) = \varnothing = \bar{\varphi}(\varnothing)$. Thus for each element $y$ of $a$ has rank less than $a$. Therefore, by the choice of $a$, for each $y$ in $a$,

$$\psi(y) = \bar{\varphi}(y).$$

For each $y$,
$$[y \in a \text{ iff } \psi(y) \in \psi(a)] \text{ and } [y \in a \text{ iff } \bar{\varphi}(y) \in \bar{\varphi}(a)].$$
Thus,
$$\psi(a) = \{\psi(y) \mid y \in a\} = \{\bar{\varphi}(y) \mid y \in a\} = \bar{\varphi}(a),$$
which is a contradiction. □

**Theorem 10.10 (Theorem 10.4)** *Let $f$ be a permutation on $A$, and let $\bar{f}$ be the extension of $f$ to $V$. Then the following nine statements hold:*

1. $\bar{f}(A) = A$
2. $\bar{f}$ is a one-to-one function on $V$.
3. $\bar{f}^{-1} = \overline{f^{-1}}$.
4. $\bar{f}$ is onto $V$.
5. For all $x$ and $y$ in $V$, $x \in y$ iff $\bar{f}(x) \in \bar{f}(y)$.
6. For all $x$ in $P$, $\bar{f}(x) = x$.
7. For all ordered $n$-tuples $(b_1, \ldots, b_n)$ of elements of $V$,
$$\bar{f}((b_1, \ldots, b_n)) = (\bar{f}(b_1), \ldots, \bar{f}(b_n)).$$
8. If $f$ is the identity function on $A$, then $\bar{f}$ is the identity function on $V$.
9. If $g$ is a permutation on $A$, then $f * g$ is a permutation on $A$ and $\bar{f} * \bar{g} = \overline{f * g}$.

**Proof.** 1. Statement 1 follows immediately from Definition 10.11.

2. $f = f_0$ is a one-to-one function on $A = V_0$. Suppose $n$ is a nonnegative integer, $f_n$ is a one-to-one function on $V_n$, and $f_n$ restricted to $V_0$, $f_n \upharpoonright V_0$, is $f_0$. Then
$$f_{n+1} \upharpoonright V_0 = f_n \upharpoonright V_0 = f_0.$$
Let $x$ and $y$ be arbitrary elements of $V_{n+1}$ such that $x \neq y$.

*Case 1.* If $x$ and $y$ are in $V_0$, then
$$f_{n+1}(x) = f_n(x) = f_0(x) \neq f_0(y) = f_n(y) = f_{n+1}(y).$$

*Case 2.* If $x$ is in $V_{n+1} - V_0$ and $y \in V_0$, then $x$ is a set and $y$ is an atom. Then by Lemma 10.2, $f_{n+1}(x) = \{f_n(u) \mid u \in x\}$ is a set and

$f_{n+1}(y) = f_0(y)$ is an atom, and therefore $f_{n+1}(x) \neq f_{n+1}(y)$. Similarly, if $y$ is in $V_{n+1} - V_0$ and $x$ is in $V_0$, then $f_{n+1}(x) \neq f_{n+1}(y)$.

Case 3. If $x$ and $y$ are in $V_{n+1} - V_0$, then, because $x \neq y$, either there exists $u$ in $x$ such that $u \notin y$ or there exists $w$ in $y$ such that $w \notin x$. Without loss of generality, let $u$ in $x$ be such that $u \notin y$. Then by Lemma 10.2,

$$f_n(u) \in \{f_n(z) \,|\, z \in x\} = f_{n+1}(x) \text{ and } f_n(u) \notin \{f_n(v) \,|\, v \in y\} = f_{n+1}(y).$$

Thus $f_{n+1}(x) \neq f_{n+1}(y)$.

Cases 1, 2, and 3 show that $f_{n+1}$ is a one-to-one function on $V_{n+1}$.

It then follows by induction it follows that $f_k$ is a one-to-one function for all $k$ in $\mathbb{I}^+$. It easily follows from this that $f_\infty$ is a one-to-one function on $V_\infty$. Repetition of the above inductive argument starting with $f_\infty$ then yields that $\bar{f}$ is a one-to-one function on $V$.

3. $f^{-1}$ is also a permutation on $A$ and it is immediate that $(f^{-1})_0 = (f_0)^{-1}$. Suppose $n$ is a nonnegative integer and $(f^{-1})_n = (f_n)^{-1}$. Let $x$ be an arbitrary element of $V_{n+1}$. If $x \in V_n$, then by hypothesis and the definitions of $f_{n+1}$ and $(f^{-1})_{n+1}$,

$$(f^{-1})_{n+1}(x) = (f^{-1})_n(x) = (f_n)^{-1}(x) = (f_{n+1})^{-1}(x).$$

If $x \in V_{n+1} - V_n$, then by Definition 10.11,

$$\begin{aligned}(f^{-1})_{n+1}[f_{n+1}(x)] &= (f^{-1})_{n+1}[\{f_n(z) \,|\, z \in x\}] \\ &= \{(f^{-1})_n[f_n(z)] \,|\, z \in x\} \\ &= \{(f_n)^{-1}[f_n(z)] \,|\, z \in x\} \\ &= \{z \,|\, z \in x\} = x,\end{aligned}$$

that is,

$$(f^{-1})_{n+1}[f_{n+1}(x)] = x. \tag{10.10}$$

Because by Statement 2, $f_{n+1}$ is a one-to-one function and $x$ is an arbitrary element of $V_{n+1}$, it follows from Equation 10.10 that

$$(f^{-1})_{n+1} = (f_{n+1})^{-1}.$$

Thus by mathematical induction $(f^{-1})_k = (f_k)^{-1}$ for all $k$ in $\mathbb{I}^+$. Therefore,

$$(f^{-1})_\infty = \bigcup_{i=0}^{\infty} (f^{-1})_i = \bigcup_{i=0}^{\infty} (f_i)^{-1} = \left(\bigcup_{i=0}^{\infty} f_i\right)^{-1} = (f_\infty)^{-1}.$$

By repeating the above induction for non-negative integers $n$ on $(f^{-1})_{\infty+n}$ and $(f_{\infty+n})$ one gets,

$$(f^{-1})_{\infty+n} = (f_{\infty+n})^{-1}.$$

Thus
$$\overline{f^{-1}} = \bigcup_{i=0}^{\infty}(f^{-1})_{\infty+i} = \bigcup_{i=0}^{\infty}(f_{\infty+i})^{-1} = \left(\bigcup_{i=0}^{\infty} f_{\infty+i}\right)^{-1} = \bar{f}^{-1}.$$

4. Let $x$ be an arbitrary element of V. Then applying Statement 2 to the permutation $f^{-1}$ on A, yields that $\overline{f^{-1}}$ is a one-to-one function on $V$. Then by Statement 3,
$$\bar{f}[\overline{f^{-1}}(x)] = \bar{f}[\bar{f}^{-1}(x)] = x,$$
which, because $x$ is an arbitrary element of V, shows that $\bar{f}$ is onto V.

5. Let $x$ and $y$ be arbitrary elements of V.
($i$) Suppose $x \in y$. Because $y$ is a set, it follows from Lemma 10.2 that
$$\bar{f}(y) = \{\bar{f}(z) \mid z \in y\} \supseteq \{\bar{f}(x)\}.$$
Thus $\bar{f}(x) \in \bar{f}(y)$.
($ii$) Suppose $\bar{f}(x) \in \bar{f}(y)$. Then by ($i$) above with $x$ replaced by $\bar{f}(x)$, $y$ by $\bar{f}(y)$, and $\bar{f}$ by $\overline{f^{-1}}$, we obtain
$$\overline{f^{-1}}[\bar{f}(x)] \in \overline{f^{-1}}[\bar{f}(y)],$$
which by $\overline{f^{-1}} = \overline{f^{-1}}$ (Statement 3) yields,
$$\bar{f}^{-1}[\bar{f}(x)] \in \bar{f}^{-1}[\bar{f}(y)],$$
from which $x \in y$ follows.

6. Suppose Statement 6 is false. A contradiction will be shown. Let $x$ be an element of P of smallest rank such that $\bar{f}(x) \neq x$. $x \neq \varnothing$, because by Definition 10.11
$$\bar{f}(\varnothing) = f_0(\varnothing) = \{f(x) \mid x \in \varnothing\} = \varnothing.$$
Let $y$ be an arbitrary element of $x$. By Theorem 8.5, $y$ is a pure set and by Theorem 8.4, $y$ has rank less than the rank of $x$. Therefore, by the choice of $x$, $\bar{f}(y) = y$. Thus, by Lemma 10.2,
$$\bar{f}(x) = \{\bar{f}(z) \mid z \in x\} = \{z \mid z \in x\} = x,$$
which contradicts the choice of $x$.

7. Statement 7 immediately follows for a 1-tuple. Suppose $x$ and $y$ are arbitrary elements of $\mathsf{V}$. Then by Lemma 10.2,

$$\begin{aligned}\bar{f}[(x,y)] &= \bar{f}[\{\{x\},\{x,y\}\}] = \{\bar{f}(\{x\}),\bar{f}(\{x,y\})\} \\ &= \{\{\bar{f}(x)\},\{\bar{f}(x),\bar{f}(y)\}\} = (\bar{f}(x),\bar{f}(y)),\end{aligned}$$

and thus Statement 7 is true for ordered 2-tuples. Because for $n > 2$,

$$(x_1,\ldots,x_n) = ((x_1,x_2,\ldots,x_{n-1}),x_n),$$

it follows by Statement 7 for ordered 2-tuples that

$$\bar{f}[(x_1,\ldots,x_n)] = (\bar{f}[(x_1,x_2,\ldots,x_{n-1})],\bar{f}[x_n]),$$

and thus by a simple argument involving mathematical induction,

$$(\bar{f}[x_1,x_2,\ldots,x_{n-1},\bar{f}[x_n]) = (\bar{f}[x_1],\ldots,\bar{f}[x_n]),$$

and thus,

$$\bar{f}[(x_1,\ldots,x_n)] = (\bar{f}[x_1],\ldots,\bar{f}[x_n]),$$

showing Statement 7.

8. Suppose $f$ is the identity function on $A$ and $\bar{f}$ is not the identity function on $\mathsf{V}$. A contradiction will be shown. Let $b$ be an element of $\mathsf{V}$ of smallest rank such that $\bar{f}(b) \neq b$. Then $b$ is not an atom, because $f$ is the identity function on $A$ and $f = \bar{f}$ on $A$ by Definition 10.11. $b$ is not $\varnothing$, because $\bar{f}(\varnothing) = \varnothing$. Thus, because by Theorem 8.4 each element of $b$ has rank $<$ the rank of $b$, it follows from the choice of $b$ and Lemma 10.2 that

$$\bar{f}(b) = \{\bar{f}(x) \,|\, x \in b\} = \{x \,|\, x \in b\} = b,$$

contradicting $\bar{f}(b) \neq b$.

9. Suppose $g$ is a permutation on $A$. Then it is immediate that $f * g$ is a permutation on $A$. Suppose and $\bar{f} * \bar{g} \neq \overline{f * g}$. A contradiction will be shown. Let $b$ be an element of $\mathsf{V}$ of smallest rank such that $\bar{f} * \bar{g}(b) \neq \overline{f * g}(b)$. Then $b$ is not an atom and $b$ is not $\varnothing$. Thus by Theorem 8.4 and the choice of $b$,

$$\bar{f} * \bar{g}(b) = \{\bar{f} * \bar{g}(x) \,|\, x \in b\} = \{\overline{f * g}(x) \,|\, x \in b\} = \overline{f * g}(b),$$

contradicting $\bar{f} * \bar{g}(b) \neq \overline{f * g}(b)$. $\square$

*Theories of Meaningfulness*

**Lemma 10.5 (Lemma 10.3)** *Suppose $D \in \mathsf{P}$ and $\varphi$ and $\psi$ are $\in$-isomorphisms from $\langle V_\infty, \in, A \rangle$ onto $\langle \mathsf{P}_D, \in', D \rangle$ (Definition 10.5). Suppose $\varphi$ and $\psi$ are also $\in$-isomorphisms of the base structure $\mathfrak{B} = \langle A, R_j \rangle_{j \in J}$ onto $\mathfrak{N} = \langle \mathsf{P}_D, S_j \rangle_{j \in J}$. Let $h$ be the restriction of $\varphi^{-1} * \psi$ to $A$. Then $\bar{h}$ is an $\in$-symmetry of $\mathfrak{B}$ and $\bar{h} \upharpoonright V_\infty = \varphi^{-1} * \psi$.*

**Proof.** It is immediate that $h$ is a permutation on $A$. Thus $\bar{h}$ is an $\in$-symmetry by Theorem 10.5. It will be shown by contradiction that

$$\bar{h} \upharpoonright V_\infty = \varphi^{-1} * \psi. \tag{10.11}$$

Suppose Equation 10.11 is false. Let $a$ in $V_\infty$ be an element of least rank such that $\bar{h}(a) \neq \varphi^{-1} * \psi(a)$. $\mathsf{rank}(a) \neq 0$, because by hypothesis, for all $b$ in $A$,

$$h(b) = \varphi^{-1} * \psi(b).$$

Thus $a$ is a set and for each element $y$ of $a$,

$$\bar{h}(y) = \varphi^{-1} * \psi(y).$$

Because $\bar{h}$ is an $\in$-symmetry, $\varphi * \bar{h}$ is an $\in$-isomorphism of $\langle V_\infty, \in, A \rangle$ onto $\langle \mathsf{P}_D, \in', D \rangle$. Therefore,

$$\begin{aligned} \varphi * \bar{h}(a) &= \{\varphi * \bar{h}(y) \mid y \in a\} \\ &= \{\varphi * \varphi^{-1} * \psi(y) \mid y \in a\} = \{\psi(y) \mid y \in a\} = \psi(a),\end{aligned}$$

and thus,

$$\bar{h}(a) = \varphi^{-1} * \psi(a),$$

a contradiction.

Let $j$ be an arbitrary element of $J$. Then, because $\varphi$ and $\psi$ are $\in$-isomorphisms from $\mathfrak{B}$ onto $\mathfrak{N}$, for each $j$ in $J$,

$$\bar{h}(R_j) = \varphi^{-1} * \psi(R_j) = \varphi^{-1}(S_j) = R_j,$$

and thus $\bar{h}$ is an $\in$-symmetry of $\mathfrak{B}$. □

## Chapter 11

# Applications, Limitations, and Generalizations of Axiom System $\mathcal{FST}$

## 11.1 AN EPISTEMOLOGY FOR A RULE BASED ON INVARIANCE

The method of "dimensional analysis" of physics is an inferential technique involving invariance. Physicists and engineers often employ it to find solutions to complicated physical problems where exact solutions by purely mathematical methods are unknown. It is also indispensable in many situations, for example airplane and ship design, where, for all practical purposes, it is impossible to provide precise and detailed formulations of the fundamental equations characterizing a needed solution.

There are two kinds of foundations for dimensional analysis: a *mathematical foundation* describing the mathematical techniques employed in dimensional analysis, and an *epistemological foundation* describing the nature of the inferential techniques employed in dimensional analysis. Measurement-theoretic versions of mathematical foundations are given in Luce (1978), Narens (2002a, chapter 4, section 10), chapter 10 of *Foundations of Measurement, Vol. I* (Krantz et al., 1971), and Section 22.7 of *Foundations of Measurement, Vol. III* (Luce et al., 1990). A full epistemological foundation is not attempted in this book. Instead a key epistemological principle of dimensional analysis is isolated and generalized. An epistemological foundation for the generalization is provided by relating scientific definability to symmetry invariance. The generalization also applies to a wide variety of nonphysical phenomena.

The following example involving a law of the pendulum nicely illustrates an elementary application of dimensional analysis.

Consider a simple pendulum consisting of a ball suspended by a string. The ball is displaced and is then released. It's period t will be found through dimensional analysis. To accomplish this, a listing is needed for the relevant physical entities that determine t. For this example this is easy:[17] they are,

- the period t,
- the mass m of the ball,
- the length d of the string,
- the gravitational constant g that describes the acceleration of a small mass towards the center of the earth,
- and the angle a that the pendulum makes with the vertical when the ball is released.

t, m, d, g, and a are qualitative entities. They are elements of the domain of a relational structure $\mathfrak{X}$ consisting of components used to measure t, m, d, g, and a; for example, a component that is an extensive structure used to measure length, a component that is an extensive structure used to measure mass, et cetera. The previously referenced measurement-theoretic foundations of dimensional analysis provide formulations for $\mathfrak{X}$ such that t, m, d, g, and a are measured in terms of a coherent set of units; for example if d is measured in terms of the unit $u$ and t in terms of the unit $v$, then g is measured in terms of the unit $u/v^2$. In this application a is measured by the ratio of a measured length of arc divided by a measured radius, and is

---

[17]In some examples, it may be tricky. Bridgman (1931) writes the following about a defense by Lord Rayleigh of a particular application of dimensional analysis:

> This reply of Lord Rayleigh is, I think likely to leave us cold. Of course we no not question the ability of Lord Rayleigh to obtain the correct result by the use of dimensional analysis, but must we have the experience and physical intuition of Lord Rayleigh to obtain the correct result also? Might not perhaps a little examination of the logic of the method of dimensional analysis enable us to tell whether temperature and heat are "really" independent units or not, and what the proper way of choosing our fundamental units is?
> 
> Besides the prime question of the proper number of units to choose in writing our dimensional formulas, this problem of heat transfer raises many others also of a physical nature. For instance, why are we justified in neglecting the density, or the viscosity, or the compressibility, or the thermal expansion of the liquid, or the absolute temperature? We will probably find ourselves able to justify the neglect of all these quantities, but the justification will involve real argument and a considerable physical experience with physical systems of the kind which we have been considering. The problem cannot be solved by the philosopher in his armchair, but the knowledge involved was gathered only by someone at some time soiling his hands with direct contact. (pp.11–12)

thus a real number—a "dimensionless" quantity—that does not depend on the unit used to measure length.

Using "physical intuition," it is assumed that the above information is sufficient for specifying the qualitative entity t; that is, it is assumed that t is physically determined by m, g, d, and a. In functional notation this is equivalent to asserting,

$$t = F(m,g,d,a). \tag{11.1}$$

A proper numerical representation for Equation 11.1 consists $(i)$ in providing proper numerical measurements, $t$, $m$, $g$, $d$, and $a$, in terms of a coherent system of units to respectively, t, m, g, d, and a, and $(ii)$ finding a numerical function $F$ such that

$$t = F(m, g, d, a). \tag{11.2}$$

A principle of dimensional analysis called *dimensional invariance* requires all other proper numerical formulations of Equation 11.1 to have the form,

$$t' = F(m', g', d', a), \tag{11.3}$$

where $t'$, $m'$, $g'$, and $d'$ are proper numerical measurements in some other coherent system of units.[18]

Note that in Equations 11.2 and 11.3, the same numerical function $F$ is used to represent the qualitative function F. $a$ also appears in Equations 11.2 and 11.3. By the way it was defined, $a$ is a real number that does not depend on which unit is used to measure length. Thus a has the same numerical value in each coherent system of units. Because t, m, and d are measured on ratio scales, $t'$, $m'$, $g'$, and $d'$ are related to $t$, $m$, $g$, and $d$ by the following: there are positive real numbers $\alpha$, $\beta$, and $\gamma$ such that

$$t' = \alpha t,\ d' = \beta d,\ g' = \frac{\beta g}{\alpha^2},\ \text{and}\ m' = \gamma m. \tag{11.4}$$

Substituting Equation 11.4 into Equation 11.3 yields,

$$\alpha t = F\left[\gamma m, \frac{\beta g}{\alpha^2}, \beta d, a\right]. \tag{11.5}$$

By dimensional invariance, Equation 11.5 is valid for all positive reals $\alpha$, $\beta$ and $\gamma$. Because the measurement of a does not vary with changes of units, Equation 11.5 can be rewritten as

$$\alpha t = H\left[\gamma m, \frac{\beta g}{\alpha^2}, \beta d\right], \tag{11.6}$$

---

[18] A theorem of Luce (1978) shows that the dimensional invariance of $F$ is equivalent to F being invariant under the symmetries of the structure $\mathfrak{X}$ described previously.

where for all positive real numbers $s$, $x$, $y$, and $w$,

$$s = F(x,y,w,a) \text{ iff } s = H(x,y,w).$$

In Equation 11.6 the first argument of $H$, $\gamma m$, can take any real value by an appropriately choosing $\gamma$ while leaving the value of $H$, $at$, unchanged. From this it follows that the value of $H$ does not depend on its first argument. Thus rewrite Equation 11.6 as,

$$at = K\left[\frac{\beta g}{\alpha^2}, \beta d\right]. \tag{11.7}$$

Choose units so that $\alpha = \sqrt{g}/\sqrt{d}$ and $\beta = 1/d$. Then by Equation 11.7,

$$t = \frac{\sqrt{d}}{\sqrt{g}} \cdot K(1,1). \tag{11.8}$$

It is easy to verify by inspection that if the measurements of **t** and **g**, and **d** are changed to another coherent system of units, then an equivalent form of Equation 11.8 is valid with the same real constant $K(1,1)$. If it is assumed that the measurement of **g** is known in one coherent set of units—and thus in all coherent set of units—then Equation 11.8 provides a method of calculating for each length **d**, the measurement of the period **t** in terms of the measurement of **d** and the real number $K(1,1)$. The number $K(1,1)$ is not completely specified by dimensional analysis. However, it can be found by experiment; that is, a particular measured length can be chosen and the period measured for that particular length and $K(1,1)$ can be computed by Equation 11.8.

In the above argument **a** is assumed to be fixed. As **a** varies so will Equation 11.8; however, Equation 11.8 will have the same form, because only the real number $K(1,1)$ will vary. In other words, $K(1,1)$ is a function of **a** and therefore of $a$. Thus for variable angle **a**, rewrite Equation 11.8 as

$$t = \frac{\sqrt{d}}{\sqrt{g}} N(a), \tag{11.9}$$

where $a$ is the (dimensionless) measurement of **a**, and $N$ is a fixed real valued function.

The usual law for the period of the pendulum derived from Newton's laws is essentially the same as Equation 11.9, but with $N(a)$ being the function $\sin(a)$. Thus by using Newton's laws, $N(a)$ can be completely determined without having to resort to experiment. However, the above dimensional analysis also yields certain completely determined laws of pendulums: For example, if two pendulums of measured lengths (in the same

## Applications, Limitations, and Generalizations of $\mathcal{FST}$

length unit) $d_1$ and $d_2$ are released from the same angle with the vertical, and their resulting periods are $t_1$ and $t_2$ (measured in the same time unit), then it is immediate from Equation 11.9 that

$$\frac{t_1}{t_2} = \frac{\sqrt{d_1}}{\sqrt{d_2}}. \qquad (11.10)$$

Note that Equation 11.10 does not depend on which units are used to measure distance and time and that it does not mention the gravitational constant.

The following principle, which is used in the above dimensional analysis argument, was recognized early in the development of physics:

> *Quantitative Dimensional Invariance*: A numerical relation that expresses a valid physical relationship between physical variables has the same mathematical form for all proper measurements of the physical variables.

In terms of the qualitative, measurement-theoretic foundation for dimensional analysis developed in Luce (1978), this principle translates into the following:

> *Qualitative Dimensional Invariance*: A qualitative relation that expresses a valid physical relationship between physical objects is invariant under the symmetries of the base structure used to measure the physical objects.

In the pendulum example, having "the same mathematical form for all proper measurements of the physical variables" of quantitative dimensional invariance is captured by the quantitative equation,

$$t = \frac{\sqrt{d}}{\sqrt{g}} \cdot K(1,1),$$

where $K(1,1)$ is a real number. In terms of qualitative invariance, this corresponds to the qualitative equation,

$$\mathsf{t} = \mathsf{F}(\mathsf{m},\mathsf{g},\mathsf{d},\mathsf{a}).$$

Thus in the pendulum example, the "same mathematical form" of quantitative dimensional analysis corresponds qualitatively to a single function. The existence of such a correspondence to a single qualitative function provides a rigorous justification for the use in this example of the phrase, "same mathematical form." However, in other applications of dimensional analysis and its generalizations, "same mathematical form" may correspond to a set of qualitative functions. In such cases, the use of the phrase "same mathematical form" appears to me to be problematic, because of a lack of a qualitative criterion for different qualitative functions to have "the same form."

As noted in the above quotation by Bridgman, in order for dimensional analysis to work properly, a listing of the relevant physical variables is needed. Qualitatively, this corresponds to describing a base structure whose numerical representation produces the correct scale families for measuring physical objects. Luce (1978) provides such a description.

In dimensional analysis, it is assumed that the desired solution $S$ to a problem satisfies Quantitative Dimensional Invariance. $S$ has a corresponding qualitative solution S, which is a higher-order relation that sometimes is a first-order relation or function. Dimensional analysis relies on $S$ being "determined by the variables." In the dimensional analysis literature, this is not a precisely defined concept. In the theories of meaningfulness developed in this part of the book, it corresponds to the precisely defined concept, "S belongs to the scientific topic specified by the base structure used to measure the physical objects." Using this version of "determined by the variables," $S$ being quantitatively dimensionally invariant then corresponds to S being invariant under the symmetries of the base structure. The base structure can be precisely described through qualitative treatments of structural properties of physical variables, like in Luce (1978), Narens (2002a, chapter 5, section 10), Krantz et al., (1971, chapter 10), and Luce et al., (1990, section 22.7).

The following two quotations of Narens formulates an epistemological rule involving the use of invariance. The rule applies to many situations outside of physics. It is extracted from principles employed in dimensional analysis. It should by no means be considered as a theory of dimensional analysis, because dimensional analysis has many other additional principles. (The quotations have been altered to refer to theorems and definitions of this book.)

> Suppose in a particular setting we are interested in finding the functional relationship of the qualitative variables, $x$, $y$, and $z$. We believe that the primitive relations (which are known) completely characterize the current situation. Furthermore, our understanding (or insight) about the situation tells us that $x$ must be a function of $y$ and $z$. (This is the typical case for an application of dimensional analysis in physics.) This unknown function—which we will call "the desired function"—must be determined by the primitives and the qualitative variables $x$, $y$, and $z$. Therefore, it should somehow be "definable" from these

> relations and variables. Even though the exact nature of the definability condition is not known, (it can be argued that) it must be weaker than the enormously powerful methods of definability encompassed by Axiom System $\mathcal{FST}$. Thus by [Theorem 10.7] we know that any function relating the variable $x$ to the variables $y$ and $z$ that is *not* invariant under the symmetries of the primitives cannot be the desired function. In many situations, this knowledge *of knowing that functions not invariant under the symmetries of the primitives cannot be the desired function* can be used to effectively find or narrow down the possibilities for the desired function. (Narens, 1988, pp. 70–71)

Narens (2002b) expanded on this theme as follows:

> Scientific inquiry is a complicated issue with many overlapping parts. I believe meaningfulness belongs primarily to the theoretical part of scientific inquiry. Because of the overlap of the theoretical part of a science with its experimental and applied parts, meaningfulness often has important ramifications in the experimental and applied parts.
>
> Meaningfulness is essentially a theoretical position about scientific content and its role in (theoretical) inference. For example, consider the case where by extra-scientific means (e.g., intuition, experience, etc.) a scientist is led to believe that a function $z = F(x, y)$ that he needs to describe from a subset of $A \times A$ into $A$ is completely determined by the observable, first-order relations $R_1, \ldots, R_n$ on $A$. Then it is reasonable for the scientist to proceed under the hypothesis that $F$ belongs to the scientific content of $\mathfrak{X} = \langle A, R_1, \ldots, R_n \rangle$, which for this discussion may be taken as the set of meaningful entities determined by Axiom System $\mathcal{FST}$. Thus the scientist assumes $F$ has a scientific definition in terms of $\mathfrak{X}$ and its primitives. By Theorem 10.7, $F$ is invariant under the symmetries of $\mathfrak{X}$. Suppose the scientist knows enough properties about $\mathfrak{X}$ and has the mathematical skill to determine the symmetry group $G$ of $\mathfrak{X}$. Then methods of analyses involving symmetries may be employed to provide information helpful in characterizing $F$. There are several methods in the literature for accomplishing this.
>
> Note that in the above process, scientific definability is used *to justify* $F$ belonging to the appropriate topic, invariance is used *as a mathematical technique* to find helpful information for characterizing $F$, and that these two uses are connected *by a theorem* of mathematical logic. Also note that the scientist's

belief that $F$ belonged to the topic generated by $\mathfrak{X}$ is extra-scientific. Therefore, the deductions based on information obtained through the above process should either be checked by experiment or be derived from accepted scientific theory and facts; i.e., they should be treated as scientific hypotheses that need corroboration. Thus, for the purposes of science, the above process is a method of generating hypotheses and not facts: If the scientist's extra-scientific beliefs are correct, then the generated hypotheses will be facts; however, the scientist has no *scientific guarantee* that his beliefs are correct. (Narens, 2002b, pp. 764–765)

Bridgman (1931) questioned the validity of the principle of dimensional invariance:

Why is it that an equation $[E]$ which correctly describes a relation between various measurable physical quantities must in its form be independent of the size of the fundamental units? There does not seem to be any necessity for this in the nature of the measuring process itself. (Bridgman, 1931, pg. 13)

Bridgman is correct in that, "There does not seem to be any necessity for this in the nature of the measuring process itself." The above discussion of dimensional analysis and the above quotations of Narens suggest that the the dimensional invariance of the equation $E$ results from the assumption that it is completely determined by the measured variables involved in it. Of course, such an assumption could be wrong for a number of reasons without invalidating "the nature of the measuring process." Thus the failure of $E$ to be dimensionally invariant need not result from a failure of principle of dimensional invariance or a failure of the measurement process, but from a failure to hypothesize a correct dependency among the various measurable physical quantities.

## 11.2 LIMITATIONS OF AXIOM SYSTEM $\mathcal{FST}$

In the middle of the 19th century, Riemann introduced new differential geometric techniques into mathematics. Riemann's geometries differed from those studied by geometers of the time in that they often had trivial transformation groups consisting of only the identity transformation. Because the points of two differential geometries are in one-to-one correspondence, all geometries resulting from Riemann's methods with trivial transformation groups were considered by the Erlanger Program to be manifestations of the same geometry. Furthermore, by the Erlanger Program, all higher-order relations on the domain of such a geometry were invariant, and thus belonged to the geometry. Therefore from the perspective of the Erlanger Program, these geometries should be separated from those with richer transformation groups and perhaps not be treated as "true geometries." However, with the introduction of the General Theory of Relativity by Einstein in 1916 such a position became untenable: It was clear that whatever a "geometry" was, Einstein's model—which at the time was the best description of physical space—was a geometry, and one that was shown to have the identity as its only symmetry. Thus the Erlanger Program applied to physical space yielded every relation based on the points of physical space being meaningful, a situation that was unacceptable to geometers and physicists of the time.

Some mathematicians tried to revive the Erlanger Program by the generalizing kinds of transformations allowed. But these failed because they produced poor concepts as to what "geometric (= meaningful)" was. The Erlanger Program quickly lost much of it influence in foundational aspects of geometry.

In measurement theory, a similar problem arises for trivial scales consisting of a single measuring function.

The equivalence of Axiom Systems $\mathcal{E}$ and $\mathcal{FST}$ presents an avenue for generalizing the Erlanger Program. Rather than trying to generalized $\mathcal{E}$ by generalizing "symmetry" and "group" (the method of the failed attempts of generalizations previously mentioned), generalize $\mathcal{FST}$ by weakening its axioms. The weakening can be accomplished in many ways. The next section describes one of these.

## 11.3 INTRINSICNESS

Narens (2002a) developed a concept called *intrinsicness* that captures a stronger form of invariance than symmetry invariance. He views intrinsicness as meaningfulness plus an additional form of invariance. He separates out two of its uses:

- instrinsicness as a theory of a class of laws,

- and intrinsicness as a way of providing a richer meaningfulness concept, for example, in situations where the base structure has only a single symmetry.

Narens (2002a) presents several theories of intrinsicness and applies them to a variety of scientific and philosophic issues. It is outside of the scope of this book to discusses them in detail. This section considers one special form of intrinsicness described in Narens (2002a). It is crafted to match the concepts developed in this part of the book. A simple example is presented involving a base structure with the identity as it's only $\in$-symmetry. Under Axiom System $\mathcal{FST}$ (or Axiom System $\mathcal{E}$—the Erlanger Program), this produces a useless meaningfulness theory where each element of $V_\infty$ is meaningful. For the situation described in the example, it is shown that intrinsicness yields a viable, alternative meaningfulness concept.

**Convention 11.1** Throughout the remainder of this section, the following conventions are observed:

- Axiom System $\mathcal{FST}$ holds with meaningfulness concept M.

- $\mathcal{F}$ is a non-empty set such that

  (i) each element of $\mathcal{F}$ is a base structure for M (Definition 10.8); and

  (ii) all elements of $\mathcal{F}$ are isomorphic.

- $\mathfrak{B} = (A, R_j)_{j \in J}$ is in $\mathcal{F}$. □

**Definition 11.1 (meaningful field of structures)** $\mathcal{F}$ in Convention 11.1 is called a *meaningful field of structures*. □

Intuitively, the base structure $\mathfrak{B}$ is a perspective that accounts for the scientific topic M under consideration. The accounting is done in terms of pure mathematics and an language appropriate to $\mathfrak{B}$. There are other base structures for M that also validly account for the scientific topic. A meaningful field $\mathcal{F}$ containing $\mathfrak{B}$ is a particular kind of set of such base structures. What makes it particular is that all elements of $\mathcal{F}$ are isomorphic to $\mathfrak{B}$. This allows for a common language, $\mathsf{L} = \mathsf{L}(\in, A, R_j, \mathfrak{B})_{j \in J}$, to be used for describing each structure in $\mathcal{F}$. The idea of intrinsicness is that some scientific entities have the same description in terms of $\mathsf{L}$ no matter which structure in $\mathcal{F}$ is being described by $\mathsf{L}$. Such entities are called "intrinsic." Intuitively, intrinsicness has an extra kind of invariance that is not necessarily captured by the symmetries of $\mathfrak{B}$.[19]

**Definition 11.2 (the language $\mathsf{L}_\mathfrak{B}$)** By definition,

$$\mathsf{L}_\mathfrak{B} = \mathsf{L}(\in, A, R_j, \mathfrak{B})_{j \in J}$$

---

[19]It is the common language and not isomorphism that is the key feature used in the concept of "intrinsicness." Thus concept of "intrinsicness" extends to situations where the structures in a family of base structures of M have a common language to account for M but are not necessarily isomorphic to one another. (See chapter 6 of Narens, 2002a.)

is the extension of the first-order language $\mathsf{L}(\in, \boldsymbol{A})$ with additional individual constant symbols $\boldsymbol{R_j}$, for $j \in J$, and $\mathfrak{B}$. □

**Convention 11.2** The individual constant symbols $\boldsymbol{R_j}$, for $j \in J$, and $\mathfrak{B}$ of $\mathsf{L}_{\mathfrak{B}}$ are interpreted for elements $\mathfrak{E}$ of $\mathcal{F}$ as follows:

- If $\mathfrak{E} = \mathfrak{B}$, then $\boldsymbol{R_j}$ is interpreted as $R_j$ and $\mathfrak{B}$ as $\mathfrak{B}$.

- If $\mathfrak{E} = \langle A, R_j^\star \rangle_{j \in J}$, then $\boldsymbol{R_j}$ is interpreted as $R_j^\star$ and $\mathfrak{B}$ as $\mathfrak{E}$. □

**Definition 11.3 ($\mathcal{F}$-intrinsicness and the set $\mathsf{I}_{\mathcal{F}}$)** Let $\mathcal{F}$ be a field of meaningful structures and $R \in \mathsf{V}$. Then $R$ is said to be $\mathcal{F}$-*intrinsic* if and only if there exists a formula $\Phi(x, x_1, \ldots, x_m)$ of $\mathsf{L}_{\mathfrak{B}}$ and pure sets $b_1, \ldots, b_m$ in $\mathsf{P}$ such that the following is a true statement about each structure in $\mathcal{F}$:

$$\Phi(R, b_1, \ldots, b_m) \wedge \forall x [\Phi(x, b_1, \ldots, b_m) \leftrightarrow x = R].$$

By definition, let

$$\mathsf{I}_{\mathcal{F}} = \{e \mid e \in \mathsf{V} \text{ and } e \text{ is } \mathcal{F}\text{-intrinsic}\}.$$

$\mathsf{I}_{\mathcal{F}}$ is called the *set of $\mathcal{F}$-intrinsic entities (of $\mathsf{M}$)*. □

If $\mathcal{F} = \{\mathfrak{B}\}$, then $\mathcal{F}$-instrinsicness becomes meaningfulness, that is, $\mathsf{I}_{\mathcal{F}} = \mathsf{M}$. Because $\{\mathfrak{B}\} \subseteq \mathcal{F}$, $\mathsf{I}_{\mathcal{F}} \subseteq \mathsf{M}$. For the case where $\mathsf{V}_\infty \subseteq \mathsf{M}$, and thus meaningfulness is of no use in drawing interesting scientific inferences, $\mathsf{I}_{\mathcal{F}}$ may be useful as an alternative meaningfulness concept. The following example illustrates this.[20]

**Example 11.1** Let

- $\mathfrak{X} = \langle A, \preceq, \oplus \rangle$ be a continuous extensive structure;

- $a$ be a fixed element of $A$;

- $\mathfrak{B} = \langle A, \preceq, \oplus, a \rangle$;

- for each $b$ in $A$, $\mathfrak{B}_b = \langle A, \preceq, \oplus, b \rangle$ (and thus $\mathfrak{B} = \mathfrak{B}_a$);

- $\mathcal{F} = \{\mathfrak{B}_b \mid b \in A\}$;

- $\mathsf{L}_{\mathfrak{B}} = \mathsf{L}(\in, \boldsymbol{A}, \preceq, \oplus, a)$;

---

[20] Other obvious methods could deal with the meaningfulness difficulties presented in the example. However, these will not work in more complicated settings in which intrinsicness produces an invariance concept that does not coincide with invariance under a group of transformations. Narens (2002a) provides an example.

- $\Phi(x, x_1, \ldots, x_m)$ be a formula of $\mathbf{L}_\mathfrak{B}$;

- $R$ be a higher-order relation on $A$ that is in $V_\infty$;

- $b_1, \ldots, b_m$ be pure sets;

- and for each $b$ in $A$, the following be a true statement about the structure $\mathfrak{B}_b$:

$$\Phi(R, b_1, \ldots, b_m) \wedge \forall x[\Phi(x, b_1, \ldots, b_m) \leftrightarrow x = R]. \qquad (11.11)$$

Note that if the symbol $\boldsymbol{a}$ occurs in $\Phi(x, x_1, \ldots, x_m)$, then in interpreting Equation 11.11 in $\mathfrak{B}_b$, the symbol $\boldsymbol{a}$ is interpreted as the element $b$ in $A$.

Let $d$ be an arbitrary element of $A$. Because by hypothesis $\mathfrak{X} = \langle A, \preceq, \oplus \rangle$ is a continuous extensive structure, it is homogeneous. Thus let $\alpha$ be an $\in$-symmetry of $\mathfrak{X}$ such that $\alpha(a) = d$. Then

(i) $\alpha$ is an $\in$-isomorphism of $\mathfrak{B}_a$ onto $\mathfrak{B}_d$.

Because for $i = 1, \ldots, m$, $b_i$ is a pure set,

(ii) $\alpha(b_i) = b_i$ for $i = 1, \ldots, m$.

(i), (ii), and Equation 11.11 imply[21] that

$$\Phi(\alpha(R), b_1, \ldots, b_m). \qquad (11.12)$$

Equations 11.11 and 11.12 imply $\alpha(R) = R$. Because $\alpha$ is an arbitrary $\in$-symmetry of $\mathfrak{X}$, it then follows that $R$ is invariant under the $\in$-symmetries of $\mathfrak{X}$.

Assume Axiom System $\mathcal{FST}$ with meaningfulness concept $\mathsf{M}$ and base structure $\mathfrak{B}$ for $\mathsf{M}$. Then, because $\mathfrak{B}$ has the identity as its only $\in$-symmetry, it follows from Theorem 10.7 that $V_\infty \subseteq \mathsf{M}$. $\mathcal{F}$, as defined in this example, is a meaningful field of structures. Let $G$ be the set of $\in$-symmetries of the continuous extensive structure $\mathfrak{X}$. The above argument shows that

$$\mathsf{I}_\mathcal{F} \cap V_\infty \subseteq \{R \mid R \in V_\infty \text{ and } \alpha(R) = R \text{ for all } \alpha \text{ in } G\}. \qquad (11.13)$$

Because $\mathfrak{X}$ is definable in terms of $A$, $\preceq$, and $\oplus$ through the sublanguage $\mathsf{L}(\in, A, \preceq, \oplus)$ of $\mathbf{L}_\mathfrak{B}$ and for each $b$ in $A$, $\preceq$ and $\oplus$ are interpreted in $\mathfrak{B}_b$ as

---

[21] The implication follows by a variant of the proof of Theorem 10.7. In the variant, well-known theorems of logic concerning isomorphisms are used in place of theorems about symmetries. The theorems state that isomorphisms preserve first-order definitions and the truth of first-order statements (e.g., see pp. 91–93 of Enderton, 1972).

respectively $\preceq$ and $\oplus$, it follows from Theorem 10.7 (taking $\mathfrak{X}$ as the base structure) that

$$\{R \mid R \in V_\infty \text{ and } \alpha(R) = R \text{ for all } \alpha \text{ in } G\} \subseteq \mathsf{I}_\mathcal{F} \cap V_\infty. \qquad (11.14)$$

Equations 11.13 and 11.14 show that $\mathsf{I}_\mathcal{F}$ is a viable alternative to M as a meaningfulness concept for $\mathfrak{B}$, because it coincides with invariance under the $\in$-symmetries of $\mathfrak{X}$ for elements of $V_\infty$. $\square$

In summary, Example 11.1 takes $\mathfrak{B}$ as the base structure for the meaningfulness concept M. Because $\mathfrak{B}$ has only a trivial $\in$-symmetry, $V_\infty \subseteq \mathsf{M}$, rendering M useless for scientific applications. By using intrinsicness, other meaningfulness concepts based on $\mathfrak{B}$ can be defined that are highly applicable. In Example 11.1, a particular field of meaningful structures $\mathcal{F}$ was constructed with $\mathfrak{B} \in \mathcal{F}$. Through intrinsicness, $\mathcal{F}$ generated the meaningfulness concept $\mathsf{I}_\mathcal{F}$. As a meaningfulness concept, $\mathsf{I}_\mathcal{F}$ is structurally identical to the restriction of the meaningfulness concept $\mathsf{M}'$ to $V_\infty$ that is generated using Axiom System $\mathcal{FST}$ and the homogeneous structure $\mathfrak{X}$ as the base structure for $\mathsf{M}'$. Obviously, for applications like dimensional analysis that draw inferences from invariance, $\mathsf{I}_\mathcal{F}$ ($= \mathsf{M}' \cap V_\infty$) is a superior meaningfulness concept to $V_\infty$ ($= \mathsf{M} \cap V_\infty$).

There are many laws in science that contain invariance beyond symmetry invariance that have formulations as intrinsic relations. The next section provides examples of a particular subclass of such laws.

## 11.4 POSSIBLE PSYCHOPHYSICAL LAWS

Luce (1959) presented a theory that was intended to generalize dimensional analysis to a wider range of scientific phenomena, particularly to phenomena in psychophysics. This theory became known as "Possible Psychophysical Laws," and it generated a considerable literature.[22]

Luce's 1959 theory is based on two principles:

> A substantive theory relating two or more variables and the measurement theories for these variables should be that:

---

[22]For the purposes of this book, the most important articles of this literature are Luce (1959, 1964), Falmagne & Narens (1983), Aczél,, Roberts, & Rosenbaum (1986), and Falmagne (2004). The articles by Luce founded the area; Falmagne & Narens (1983) and Aczél et al. (1986) presented better formulations of the theory under weakened assumptions. Aczél et al. (1986) provided considerable insight into the mathematics that produces laws that are single functions. Falmagne & Narens (1983) provided an important generalization of the theory to laws that are families of functions, and Falmagne (2004) applied this generalization to several physical laws.

1. [Consistency Principle] (*Consistency of substantive and measurement theories*) Admissible transformations of one or more of the independent variables shall lead, via the substantive theory, only to admissible transformations of the dependent variables.

2. [**Invariance Principle**] (*Invariance of the substantive theory*) Except for the numerical values of parameters that reflect the effect on the dependent variables of admissible transformations of the independent variables, the mathematical structure of the substantive theory shall be independent of admissible transformations of the independent variables. (Luce, 1959, pg. 85)

In working out specific functional relationships, Luce restricted himself to the case of an unknown function of a single independent variable. He also assumed that both the dependent and independent variables of this relationship were measurable by some combination of ratio, interval, and log-interval scales, and that both variables range over continua. The following is typical of his method of reasoning about this type of situation:

> Let $x \geq 0$ denote a typical value of the independent variable and $u(x) \geq 0$ the corresponding value of the dependent variable, where $u$ is the unknown functional law relating them. Suppose, first, that both variables form ratio scales. If the unit of the independent variable is changed by multiplying all values by a positive constant $k$, then according to the principle stated above only an admissible transformation of the dependent variable, namely multiplication by a positive constant, should result and the form of the functional law should be unaffected. That is to say, the changed unit of the dependent variable may depend upon $k$, but it shall not depend upon $x$, so we denote it by $K(k)$. Casting this into mathematical terms, we obtain the functional equation
> $$u(kx) = K(k)u(x)$$
> where $k > 0$ and $K(k) > 0$. (Luce, 1959, pg. 86)

By similar lines of reasoning, functional equations for other combinations of scale types of variables are arrived at, and these are summarized in Table 11.1.

Luce showed that the solutions to these functional equations are highly constrained and yield the "possible laws" in Table 11.2.

Rozeboom (1962) showed that Luce's theory was too specific, and Luce (1962) agreed that it was too restrictive to capture all possible psychophysical laws.

Applications, Limitations, and Generalizations of $\mathcal{FST}$

| SCALE TYPES | | Functional Equation | Comments |
|---|---|---|---|
| Independent Variable | Dependent Variable | | |
| ratio | ratio | $u(kx) = K(k)u(x)$ | $k > 0$, $K(k) > 0$ |
| ratio | interval | $u(kx) = K(k)u(x) + C(k)$ | $k > 0$, $K(k) > 0$ |
| ratio | log-interval | $u(kx) = K(k)u(x)^{C(k)}$ | $k > 0$, $K(k) > 0$, $C(k) > 0$ |
| interval | ratio | $u(kx + c) = K(k,c)u(x)$ | $k > 0$, $K(k,c) > 0$ |
| interval | interval | $u(kx + c) = K(k,c)u(x)$ $+ C(k,c)$ | $k > 0$, $K(k,c) > 0$ |
| interval | log-interval | $u(kx + c) =$ $K(k,c) \cdot u(x)^{C(k,c)}$ | $k > 0$, $(k,c) > 0$, $C(k,c) > 0$ |
| log-interval | ratio | $u(kx^c) = K(k,c)u(x)$ | $k > 0$, $c > 0$, $K(k,c) > 0$ |
| log-interval | interval | $u(kx^c) = K(k,c)u(x)$ $+ C(k,c)$ | $k > 0$, $c > 0$, $K(k,c) > 0$ |
| log-interval | log-interval | $u(kx^c) = K(k,c)u(x)^{C(k,c)}$ | $k > 0$, $c > 0$, $K(k,c) > 0$, $C(k,c) > 0$ |

Table 11.1: The Functional Equations for the Laws Satisfying the Principle of Theory Construction

| SCALE TYPES | | Functional Equation | Comments |
|---|---|---|---|
| Independent Variable | Dependent Variable | | |
| ratio | ratio | $u(x) = \alpha x^\beta$ | $\beta/x$; $\beta/u$ |
| ratio | interval | $u(x) = \alpha \log x + b$, $u(x) = \alpha x^\beta + \delta$ | $\alpha/x$ $\beta/x$; $\beta/u$; $\delta/x$ |
| ratio | log-interval | $u(x) = \delta e^{\alpha x^\beta}$ | $\alpha/u$; $\beta/x$; $\beta/u$; $\delta/x$ |
| ratio | ratio | $u(x) = \alpha x^\beta$ | $\beta/x$; $\beta/u$ |
| interval | ratio | **impossible** | |
| interval | interval | $u(x) = \alpha x + \beta$ | $\beta/x$ |
| interval | log-interval | $u(x) = \alpha e^{\beta x}$ | $\alpha/x$; $\beta/u$ |
| log-interval | ratio | **impossible** | |
| log-interval | interval | $u(x) = \alpha \log x + \beta$ | $\alpha/x$ |
| log-interval | log-interval | $u(x) = \alpha x^\beta$ | $\beta/x$; $\beta/u$ |

Table 11.2: The Possible Laws Satisfying the Principle of Theory Construction (*Note:* The notation $\alpha/x$ means "$\alpha$ is independent of the unit of $x$.")

Although restrictive, Luce's general approach is still very useful. One potential problem from a measurement standpoint of Luce's approach is that the admissible transformations mentioned in the Consistency Principle may be different from the admissible transformations mentioned in the Invariance Principle. This is because the Invariance Principle adds restrictions on how the independent variables are to be measured that are not contained in the Consistency Principle. This is most easily seen by considering a qualitative formulation of the Principles.

Luce assumed that the scales measuring the independent and dependent variables were given. Qualitatively this amounts to assuming a qualitative structure of the form,

$$\mathfrak{M} = \langle A, X, Y, \preceq_X, \preceq_Y, R_j, S_k \rangle_{j \in J; k \in K},$$

where $\preceq_X$ and $R_j$, $j \in J$, are relations on $X$, and $\preceq_Y$ and $S_k$, $k \in K$, are relations on $Y$, and

- $\mathfrak{X} = \langle X, \preceq_X, R_j \rangle_{j \in J}$ is the continuous qualitative structure to measure the independent variable,

- $\mathfrak{Y} = \langle Y, \preceq_Y, S_k \rangle_{k \in K}$ is the continuous qualitative structure to measure the dependent variable,

- and $A = X \cup Y$ and $X \cap Y = \varnothing$.

Let

- $G$ be the symmetry group of $\mathfrak{M}$,

- $G_X$ the symmetry group of $\mathfrak{X}$,

- and $G_Y$ the symmetry group of $\mathfrak{Y}$.

For each $\alpha$ in $G_X$ and each $\beta$ in $G_Y$, let $\mu_{\alpha,\beta}$ be the function on $A$ such that for all $a$ in $A$,

$$\mu_{\alpha,\beta}(a) = \begin{cases} \alpha(a) & \text{if } a \in X \\ \beta(a) & \text{if } a \in Y. \end{cases}$$

It is not difficult to sow that

$$G = \{\mu_{\alpha,\beta} \mid \alpha \in G_X \text{ and } \beta \in G_Y\}.$$

$G_X$ and $G_Y$ correspond to scales $\mathcal{S}_X$ and $\mathcal{S}_Y$ used to measure, respectively, the independent and dependent variables of a quantitative object $f$ that Luce calls a "function." In the qualitative generalization of Luce's theory, $f$ is allowed to correspond to one of two kinds of entities: a strictly increasing function $F$ from $\langle X, \preceq_X \rangle$ onto $\langle Y, \preceq_Y \rangle$, or a non-empty set $\mathcal{H}$ of $\preceq$-strictly increasing functions from $\langle X, \preceq_X \rangle$ onto $\langle Y, \preceq_Y \rangle$.

Luce's principal intended application was psychophysical laws. The natural interpretation of this is that $(i)$ each stimulus $x$ in $X$ gives rise a unique sensation $y$ in $Y$ in the observer, and $(ii)$ this is best modeled by a single function $F$ defined by $F(x) = y$. Thus the psychophysical situation is described by the structure

$$\mathfrak{Z} = \langle A, X, Y, \preceq_X, \preceq_Y, R_j, S_k, F \rangle_{j \in J;\, k \in K} \,.$$

*Because the function $F$ has been added to the primitives of $\mathfrak{M}$ to form the structure $\mathfrak{Z}$, the symmetry group of $\mathfrak{M}$ is not necessarily the symmetry group of $\mathfrak{Z}$.* A consequence of this is that $\mathfrak{M}$ may no longer be appropriate to measure $X$ and $Y$ in applications involving $\mathfrak{Z}$. In particular, the inclusion of $F$ as part of the measurement process makes some of the entries in Luce's tables invalid. Narens (2002a) writes the following about this:

> For concreteness, we will consider the case where the dependent variable is a ratio scale and the independent variable is an interval scale. Similar conclusions for the other cases follow by similar arguments. In order to discuss qualitative as well as quantitative issues, we will suppose that the independent variable results from measurement of the qualitative structure $\mathfrak{X}$ by the ratio scale $\mathcal{S}$ onto $\mathbb{R}^+$, the dependent variable results from measurement of the qualitative structure $\mathfrak{Y}$ by interval scale $\mathcal{T}$ onto $\mathbb{R}$, and the domains $X$ of $\mathfrak{X}$ and $Y$ of $\mathfrak{Y}$ are disjoint. Let $f$ be a function between the measurements of $\mathfrak{X}$ and $\mathfrak{Y}$ that is a "law" relating $\mathfrak{X}$ and $\mathfrak{Y}$ in sense of Luce (1959), that is, let $\varphi \in \mathcal{S}$, $\psi \in \mathcal{T}$, and $f$ be a continuous function from $\mathbb{R}^+$ onto $\mathbb{R}$, and let the following condition be satisfied:
>
> > For each $r \in \mathbb{R}^+$ there exists $\psi'$ in $\mathcal{T}$ such that for each $x$ in $X$,
> >
> > $$\psi[f(r\varphi(x))] = \psi'[f(\varphi(x))] \,.$$
>
> Under the above conditions, Luce (1959) showed that $f$ has [a solution of] the following form: There exist [real $a$ and $b$, $a \neq 0$] such that for all $x$ in $X$,
>
> $$f(\varphi(x)) = a \log(\varphi(x)) + b \,. \qquad (11.15)$$
>
> A consequence of this is that $f$ is a one-to-one function. Under any reasonable concept of "law" it appears to be eminently

reasonable that if a one-to-one function is a "law", then its inverse should also be a "law". However, by Luce's 1959 theory this need not be the case, and in fact by Luce's theory it is not the case for many important situations like the case of the "law" represented by the function $f$ above. The reason for this is that the "inverse" of the above "law" would be a "law" with the independent variable an interval scale and the dependent variable a ratio scale—one of the impossible cases for a "law" of Luce (1959). This failure about inverses suggests that there is something amiss about Luce's method of obtaining possible psychophysical laws. One way to investigate the nature of this problem is through meaningfulness considerations. (pp. 254–255)

Narens (2002a) shows that if

- $\mathfrak{X}$ is measured on a ratio scale family of isomorphisms $\mathcal{S}$,
- $\mathfrak{Y}$ is measured on an interval scale family of isomorphisms $\mathcal{T}$,
- and for all $\alpha$ in the symmetry group of $\mathfrak{X}$ there exists $\beta$ in the symmetry group of $\mathfrak{Y}$ such that for all $x$ in $A$,

$$\beta(F(x)) = F(\alpha(x)),$$

then $\mathfrak{Z}$ is properly measured by a scale $\mathcal{Z}$ of isomorphisms consisting of all functions of the form $\varphi \cup \psi$ where

- $\varphi \in \mathcal{S}$,
- and $\psi \in \mathcal{T}'$, where $\mathcal{T}'$ is the subscale of $\mathcal{T}$ that is a translation scale, that is, where

$$\mathcal{T}' = \{\psi + s \,|\, s \in \mathbb{R}\}.$$

Notice that contrary to the assumption of Luce's condition, $\mathcal{T}$ is not the scale family used to measure the dependent variable of $F$. Instead, only the part of $\mathcal{T}$ that is a translation scale is used to measure the dependent variable of $F$. In other words, it is not proper to use the interval scale $\mathcal{T}$ to measure the dependent variable of $F$, *when $F$ is considered part of the structure* $\mathfrak{Z}$. It is proper to measure the structure $\mathfrak{Y}$ by $\mathcal{T}$, but such measurement is not measurement of the dependent variable of $F$, because $\mathfrak{Y}$ by itself does not contain enough structure to be able to "say" that the elements of its domain are the values of the dependent variable of $F$.

Luce (1959) only considers the case of a function of a single independent variable. The situation of a function of many independent variables is far richer and more interesting (e.g., see Aczél et al., 1986), and the generalization to higher-order situations involving families of functions is even more richer (see Falmagne & Narens, 1983; Falmagne, 2004). All these situations involve variants of the principles of Consistency of Substantive and Measurement Theories and the Invariance of the Substantive Theory. Narens (2002a) shows that when viewed qualitatively, the generalized theories are about qualitative functional relationships among the elements of the symmetry groups of the structures used to measure the variables. Because different sets of primitives can specify the same symmetry group, the resulting generalized possible psychophysical laws do not depend on a particular set of primitives used to the measure the variables. Because of this, the generalized possible pyschophysical laws are a special case of an intrinsicness concept developed in Narens (2002a). Viewed this way, the generalized possible psychophysical laws are not only meaningful, but often have an extra kind of invariance that is not captured by the symmetry groups of the measurement structures of variables, and thus deserve to be labeled as "laws."

## 11.5 DISTINGUISHING EMPIRICAL AND MEANINGFUL RELATIONS

In the measurement and meaningfulness literatures, meaningfulness has sometimes been interpreted as form of empiricalness. I see this as a mistake. "Empiricalness" refers to the verification or refutability of propositions about real-world phenomena, and is often associated with concepts of error. Even if "empirical" is idealized to infinite situations in which error plays no role, which is often the case in scientific-philosophical discussions, empiricalness and meaningfulness are still about very different things. Narens (2002a) says the following about distinguishing the concepts "empirical" and "meaningful":

> the methods for establishing the empiricalness of, for example, a relation $T$ on the domain of an empirical structure $\mathfrak{E} = \langle E, E_1, \ldots, E_n \rangle$ usually depends in part on relations and methods not contained in the [scientific] topic determined by $\mathfrak{E}$. For example, consider the case where $E$ is a set of physical stimuli, for example, lights of different frequencies and intensities, $E_1, \ldots, E_n$ are behavioral relations based on $E$, involving a subject's behavior in a psychological experiment, and $\mathfrak{E} = \langle E, E_1, \ldots, E_n \rangle$. Then empirical relations and processes *from physics* may be used freely in establishing the empiricalness of a relation $H$ on $E$. These auxiliary empirical physical

relations need not be based on $E$, but can come from parts of physics that are not exclusively concerned with the characterization of physical lights. Suppose $\mathfrak{E}' = \langle E', E'_1, \ldots, E'_n \rangle$ is another empirical structure, and suppose $f$ is an isomorphism of $\mathfrak{E}$ onto $\mathfrak{E}'$. Note that the existence of the isomorphism $f$ between $\mathfrak{E}$ and $\mathfrak{E}'$ does not imply that the *empirical* physical environment in which $\mathfrak{E}$ is imbedded has an isomorphic *empirical* counterpart in which $\mathfrak{E}'$ is imbedded. Thus $f(H)$ may not be empirical.

The main point of the previous example is that for defining or constructing a particular empirical relation on the domain $E$, *any other* empirical relation on $E$, *or on other domains,* may be used in the defining or the constructing. This is a key characteristic of the concept "empirical" that is in wide variance with properties of the concept[s] "qualitative" [and "meaningful"].

It appears to me that when authors want to talk about an empirical situation [$\mathfrak{E} = \langle E, E_1, \ldots, E_n \rangle$], they usually have in mind those relations and statements that are empirical *and* are formulated in terms of the primitives of [$\mathfrak{E}$ using] an appropriate higher-order language. (pp. 409–411)

# References

Aczél, J., & Roberts, F. S. (1989). On the possible merging functions. *Mathematical Social Science, 17,* 205–243.

Aczél, J., Roberts, F. S., & Rosenbaum, Z. (1986). On scientific laws without dimensional constants. *Journal of Mathematical Analysis and Applications, 119,* 389–416.

Alper, T. M. (1985). A note on real measurement structures of scale type $(m, m+1)$. *Journal of Mathematical Psychology, 31,* 135–154.

Alper, T. M. (1987). A classification of all order-preserving homomorphism groups of the reals that satisfy finite uniqueness. *Journal of Mathematical Psychology, 31,* 135–154.

Birnbaum, M. H. (1978). Differences and ratios in psychophysical measurement. In N. J. Castellan, Jr. & F. Restle (Eds.), *Cognitive theory (Vol. 3)* (pp. 33–74). Hillsdale, NJ: Lawrence Erlbaum Associates.

Birnbaum, M. H. (1982). Controversies in psychological measurement. In B. Wegener (Ed.), *Social attitudes and psychological measurement* pp. 401–485). Hillsdale, NJ: Lawrence Erlbaum Associates.

Bridgman, P. (1931). *Dimensional analysis.* New Haven, CT: Yale University Press.

Cantor, G. (1895). Beiträge zur Begründung der transfiniten Mengenlehre. *Mathematische Annalen, 46,* 481–512.

Dzhafarov, E. N., & Colonius, H. (2001). Multidimensional Fechnerian scaling: Basics. *Journal of Mathematical Psychology 45,* 670–719.

Ekman, G. (1962). Measurement of moral judgments: A comparison of scaling methods. *Perceptual and Motor Skills, 15,* 3–9.

Ekman, G., & Künnapas, T. (1962a). Scales of aesthetic value. *Perceptual and Motor Skills, 14,* 19–26.

Ekman, G., & Künnapas, T. (1962b). Measurement of aesthetic value by "direct" and "indirect" methods. *Scandinavian Journal of Psychology, 3*, 33–39.

Ellermeier, W., & Faulhammer, G. (2000). Empirical evaluation of axioms fundamental to Stevenss ratio-scaling approach: I. Loudness production. *Perception & Psychophysics, 62(8)*, 1505–1511.

Ellermeier, W., Narens, L., & Dielmann, B. (2003). Perceptual ratios, differences, and the underlying scale. In B. Berglund and E. Borg (Eds.), *Fechner Day 2003. Proceedings of the 19th annual meeting of the International Society for Psychophysics* (pp. 71–76). Stockholm, Sweden: International Society for Psychophysics.

Enderton, H. B. (1972). *A Mathematical introduction to logic*. New York: Academic Press.

Falmagne, J.-C. (1985). *Elements of psychophysical theory*. New York: Oxford University Press.

Falmagne, J.-C. (2004). Meaningfulness and order-invariance: Two fundamental principles for scientific laws. *Foundations of Physics, 34*, 1341–1348.

Falmagne, J.-C., & Narens, L. (1983). Scales and meaningfulness of quantitative laws. *Synthese, 55*, 288–325.

Fechner, G. T. (1860). *Elemente der Psychophysik*. Amsterdam: E. J. Bonset.

Garner, W. R. (1954). A technique and a scale for loudness measurement. *Journal of the Acoustical Society of America, 26*, 73–88.

Hagerty, M., & Birnbaum, M. H. (1978). Nonmetric tests of ratio vs. subtractive theories of stimulus comparison. *Perception & Psychophysics, 24*, 121–129.

Helmholtz, H. v. (1887). Zählen und Messen erkenntnistheoretisch betrachtet. *Philosophische Aufsätze Eduard Zeller gewidmet*. Leipzig.

Hölder, O. (1901). Die Axiome der Quantität und die Lehre vom Mass [The axioms of quantity and the theory of measurement]. *Berichte über die Verhandlungen der Königlich Sächsischen Gesellschaft der Wissenschaften zu Leipzig, Mathematisch-Physikalische Classe, Bd. 53*, 1–64. (Part I translated into English by J. Michell and C. Ernst, *Journal of Mathematical Psychology, 1996, Vol. 40*, 235–252.)

Klein, F. (1872). Vergleichende Betrachtungen über neuere geometrische Vorschungen, Programm zu Eintritt in die philosophische Facultät und den Senat der Universität zu Erlangen, Erlangen, Deichert.

# References

Krantz, D. H. (1972). A theory of magnitude estimation and crossmodality matching. *Journal of Mathematical Psychology 9*, 168–199.

Krantz, D. H., Luce, R. D., Suppes, P., & Tversky, A. (1971). *Foundations of measurement, Vol. I.* New York: Academic Press.

Luce, R. D. (1959). On the possible psychophysical laws. *Psychological Review, 66*, 81–95.

Luce, R. D. (1962). Comments on Rozeboom's criticisms of "On the possible psychophysical laws." *Psychological Review, 69*, 548–551.

Luce, R. D. (1964). A generalization of a theorem of dimensional analysis. *Journal of Mathematical Psychology, 1*, 278–284.

Luce, R. D. (1978). Dimensionally invariant laws correspond to meaningful qualitative relations. *Philosophy of Science, 45*, 1–16.

Luce, R. D. (1990). "On the possible psychophysical laws" revisited: remarks on cross-modal matching. *Psychological Review, 97*, 66–77.

Luce, R. D., & Edwards, W. (1958). The derivation of subjective scales from just noticeable differences. *Psychological Review, 65*, 222–37.

Luce, R. D., Krantz, D. H., Suppes, P., & Tversky, A. (1990). *Foundations of measurement, Vol. III.* New York: Academic Press.

Luce, R. D., & Narens, L. (1985). Classification of concatenation structures by scale type. *Journal of Mathematical Psychology, 29*, 1–72.

Marley, A. A. J. (1972). Internal state models for magnitude estimation and related experiments. *Journal of Mathematical Psychology 9*, 306–319.

Miyamoto, J. M. (1983). An axiomatization of the ratio/difference representation. *Journal of Mathematical Psychology, 27*, 439–455.

Narens, L. (1981a). A general theory of ratio scalability with remarks about the measurement-theoretic concept of meaningfulness. *Theory and Decision, 13,*, 1–70.

Narens, L. (1981b). On the scales of measurement. *Journal of Mathematical Psychology, 24*, 249–275.

Narens, L. (1985). *Abstract measurement theory.* Cambridge, Ma: The MIT Press.

Narens, L. (1988). Meaningfulness and the Erlanger Program of Felix Klein. *Mathématiques Informatique et Sciences Humaines, 101*, 61–72.

Narens, L. (1994). The measurement theory of dense threshold structures. *Journal of Mathematical Psychology, 38*, 301–321.

Narens, L. (1996). A theory of magnitude estimation. *Journal of Mathematical Psychology, 40*, 109–129.

Narens, L. (1997). On subjective intensity and its measurement. In A. A. J. Marley (Ed.), *Choice, decision, and measurement: Essays in honor of R. Duncan Luce* (pp. 189–206). Mahwah, NJ: Lawrence Erlbaum Associates.

Narens, L. (2002a). *Theories of meaningfulness*. Mahwah, NJ: Lawrence Erlbaum Associates.

Narens, L. (2002b). A meaningful justification for the representational theory of measurement. *Journal of Mathematical Psychology, 46*, 746–768.

Narens, L. (2006). Symmetry, Direct Measurement, and Torgerson's Conjecture. *Journal of Mathematical Psychology, 50*, 290–301.

Narens, L., & Luce, R. D. (1983). How we may have been misled into believing in the intercomparability of utility. *Theory and Decision, 15*, 247–260.

Narens, L., & Mausfeld, R. (1992). On the relationship of the psychological and the physical in psychophysics. *Psychological Review, 99*, 467–479.

Peißner, M. (1999). *Experimente zur direkten Skalierbarkeit von gesehenen Helligkeiten*. Unpublished master's thesis. Regensburg: Universität Regensburg.

Pfanzagl, J. (1968). *Theory of measurement*. New York: Wiley.

Plateau, M. J. (1872). Sur la measure des sensations physiques, et sur la loi qui lie l'intensité de ces sensations à l'intensité de la cause excitante. *Bulletin de l'Académie Royale des Sciences, es Lettres et des Beaux Arts de Belgique, 33*, 376–388.

Pollatsek, A. and Tversky, A. (1970). A theory of risk. *Journal of Mathematical Psychology, 7*, 540–553.

Roberts, F. S. (1985). Applications of the theory of meaningfulness to psychology. *Journal of Mathematical Psychology, 29*, 311–332.

Roskam, E. E. (1989). Formal models and axiomatic measurement. In E. E. Roskam (Ed.), *Mathematical psychology in progress* (pp. 49- 68). Berlin: Springer-Verlag.

Rozeboom, W. W. (1962). The untenability of Luce's principle. *Psychological Review, 69*, 542–547.

Schneider, B. A. (1980). A technique for the nonmetric analysis of paired comparisons of psychological intervals. *Psychometrika, 45*, 357–372.

Scott, D., & Suppes, P. (1958). Foundational aspects of theories of measurement. *Journal of Symbolic Logic, 23*, 113–128.

Shepard, R. N. (1978). On the status of "direct" psychological measurement. In C. W. Savage (Ed.), *Minnesota studies in the philosophy of science (Vol. 9)* (pp. 441–490). Minneapolis, MN: University of Minnesota Press.

Shepard, R. N. (1981). Psychological relations and psychophysical scales: On the status of "direct" psychophysical measurement. *Journal of Mathematical Psychology 24*, 21–57.

Steingrimsson, R., & Luce, R. D. (2005). Evaluating a model of global psychophysical judgments: II. Behavioral properties linking summations and productions. *Journal of Mathematical Psychology, 49*, 308–319.

Stevens, S. S. (1946). On the theory of scales of measurement. *Science, 103*, 677–680.

Stevens, S. S. (1951). Mathematics, measurement and psychophysics. In S. S. Stevens (Ed.), *Handbook of experimental psychology* (pp. 1–49). New York: Wiley.

Stevens, S. S., & Davis, H. (1938). *Hearing*. New York: Wiley.

Suppes, P., Krantz, D. H., Luce R. D., & Tversky, A. (1990). *Foundations of measurement, Vol. II*. New York: Academic Press.

Suppes, P., & Zinnes, J. (1963). Basic measurement theory. In R. D. Luce, R. R. Bush, & E. Galanter (Eds.), *Handbook of mathematical psychology, Vol. 1* (pp. 1–76). New York: Wiley.

Torgerson, W. S. (1961). Distances and ratios in psychological scaling. *Acta Psychologica, 19*, 201–205.

Veit, C. T. (1978). Ratio and subtractive processes in psychophysical judgment. *Journal of Experimental Psychology: General, 107*, 81–107.

Zimmer, K. (2005). Examining the validity of numerical ratios in loudness fractionation. *Perception & Psychophysics, 67*, 569–579.

# Index

$(u, v)$, 18
$A$, 105, 106
$D_{[n]}$, 122
$D_\infty$, 122
$P_\infty$, 109
$P_n$, 109
$V_\infty$, 109
$V_n$, 109
$*$, 18
$\bar{f}$, 127
$\mathbb{I}$, 18
$\mathbb{I}^+$, 18
$\mathbb{R}$, 4, 18
$\mathbb{R}^+$, 4, 18
$\Delta$, 119
$\mathcal{FST}$, 126
$\mathcal{ST}(\mathfrak{B})$, 126
$\varnothing$, 18
$\iota_r$, 49
$\in \restriction S$, 103
$\in$, 18
$\in$-isomorphism
    from $\mathfrak{B}$ onto $\mathfrak{N}$, 125
$\in$-symmetry, 129
    of $\mathfrak{B}$, 129
$\bar{p}$, 66
$\prec$, 21
$\preceq$, 21
$\preceq$-strictly increasing, 21
$\subset$, 18
$\subseteq$, 18
$\mathsf{I}_\mathcal{F}$, 153

$\mathsf{L}_\mathfrak{B}$, 152
$\restriction$, 108
$\wp(S)$, 106
$n$-ary relation, 108
$\mathsf{L}(\in, A)$, 101–105
$\mathcal{E}$, 131
$\mathsf{N}$, 119
$\mathsf{M}$, 121
$\mathsf{P}$, 109
$\mathsf{V}$, 109
$\mathrm{rank}(x)$, 110
$\varphi(\mathrm{R})$, 19

## A

Aczél, J., 93, 155, 161
Alper, T., 31
Alper, T. M., 45
analytic geometry, 40
antisymmetric, 21
associative, 25
atom, 105
averaging of rating data, 92–93
Axiom of Measurement, 122

## B

base structure, 119
    for M, 126
Birnbaum, M. H., 80
bisymmetry, 25
        bisection structure, 24

Bridgman, P., 150

## C

Cantor, G., 21, 22
Cauchy, 11
Cauchy's Equation, 11
codomain of $R$, 108
Colonius, H., 56
commutativity
    bisection structure, 24
commute, 37
comprehension principle, 106
connected, 21
continuous bisection structure, 24
continuous extensive structure, 25
continuous ratio production, 69–70
    axioms for, 69
continuous structure, 23
continuous threshold structure, 48–52
    representation theorem, 48
continuum, 21

## D

Dedekind complete, 21
dense, 25
denumerable, 21
denumerable density, 22
Dielmann, B., 79
dimensional analysis, 143–150
dimensional invariance, 145
    qualitative, 147
    quantitative, 147
domain, 18
domain of $R$, 108
Dzhafarov, E. N., 56

## E

Edwards, W., 56

Ekman, G., 74
Ellermeier, W., 68, 79
empirical, 161–162
endomorphism, 97
entity, 105
    purely mathematical, 109
    scientific, 109
Erlanger Program, 40–45, 127, 135, 151
    comparison with measurement theory, 43–45
existence theorem, 22
extension of $f$ to $\mathsf{V}$, 128
extensionality principle, 106

## F

Falmagne, J.-C., 10, 155, 161
Faulhammer, G., 68
Fechner, 55
Fechner structure, 48
    $\infty$-point unique, 50
    homogeneous, 49
    translation by $r$, 49
Fechner, G. T., 2, 56
first-order language, 101
first-order relation, 112
formulas (of $\mathsf{L}(\in, \mathbf{A})$), 102
Full Definability, 126
Full Scientific Topic, 126
Full Use of Pure Mathematics, 126
function, 108

## G

Garner, W. R., 74, 81
group
    $\infty$-point homogeneous, 37
    $\infty$-point unique, 37
    $m$-point homogeneous, 36
    $n$-point unique, 37

1-point homogeneous, 37
commutative, 37
invariant of, 41
of symmetries, 36
of transformations, 36

## H

Hölder, O., 25
Hagerty, M., 81
Helmholtz, H. v., 25
higher-order, 112
homomorphism
   textbf, 95

## I

idempotence
   bisection structure, 24
indirect method, 47
induced discrimination relation, 56
interpersonal comparison, 92
interval scale
   subscale, 31
intrinsic, 151–155
   $\mathcal{F}$-intrinsic, 153
invariant, 41
isomorphism, 19

## K

Künnapas, T., 74
Klein, F., 41
Krantz, D. H., 1, 143, 148

## L

language $\mathsf{L}(\in, \mathbf{A})$, 101
Logarithmic Law, 55
Lord Rayleigh, 144

Luce, R. D., 1, 45, 56, 68, 93, 143, 148, 155

## M

magnitude estimation, 63
magnitude measurement
   axioms for, 64–68
   behavioral structure, 66
   Commutative Axiom, 67
   Multiplicative Axiom, 68
   multiplicative measuring
      function, 67
   multiplicative property, 63
   Steven's measuring function, 66
   Stevens scale, 66
magnitude production, 61–71
   ratio, 62
Mausfeld, R., 59
meaningful, 17, 115
   $\mathcal{S}$-representationally, 33
   distinguished from "empirical",
      161–162
   perceived risk, 89
   rating data, 92
meaningful field of structures, 152
meaningfulness concept, 126
meaningless, 17, 115
   $\mathcal{S}$-representationally, 33
measuring function, 22
monotonicity
   bisection structure, 24, 75, 82
   extensive structure, 25

## N

Narens, L., 1, 17, 30, 31, 45, 48, 52, 59, 64, 78, 79, 93, 98, 143, 148–151, 155, 161
non-set, 105

## O

ordered $n$-tuple, 107

## P

Peißner, M., 68
permutation on $A$, 127
Pfangzagl, J., 95
Pfanzagl, J., 1, 45, 74
Plateau, M. J., 2–4, 8–10, 24, 25
    his power law, 8
    his theory, 2–4
        issues generated by, 8–10
Pollatsek, A., 89
positivity, 26
possible psychophysical
        laws, 155–161
power set, 106
Preserved Equal Ratios Law, 5
Preserved Midway Ratio Law, 3, 8
primitives, 18
psychological primitives, 56
psychological structure, 57
psychophysical function, 2, 4
Psychophysical Power Law, 5, 8
pure mathematics, 117–118
purely mathematical entity, 109

## Q

qualitatively $\mathcal{S}$-meaningful, 95
quantitatively $\mathcal{S}$-meaningful, 89

## R

rank, 110
rank function, 109–111
relation
    0-ary, 18
    1-ary, 18

relational structure, 18
representation, 24
representational theory, 23
restriction of $\in$ to $S$, 103
restriction of $f$ to $S$, 108
Roberts, F. S., 93, 95, 155
Rosenbaum, Z., 155
Roskam, E., 90
Rozeboom, W. W., 156

## S

scale, 22
    $\infty$-point homogeneous, 30
    $\infty$-point unique, 30
    $m$-point homogeneous, 30
    $n$-point unique, 30
    1-point homogeneous, 30
    2-point homogeneous, 30
    absolute, 32
    homogeneous, 29, 31
    interval, 23
    log-interval, 28
    nominal, 32
    ordinal, 23
    ratio, 23
    translation, 28
scale family, 22
    representationally
        equivalent, 29
scientific definability
    axiom of, 126
scientific definition, 119
scientific domain, 106
scientific entity, 109
scientific topic, 115–120
    principles, 115–116
Scott, D., 44, 94
sentences (of $\mathbf{L}(\in, \boldsymbol{A})$), 103
set, 105
set model, 104
solvability

bisection structure, 24
    extensive structure, 26
Steingrimmsson, R., 68
Stevens, S. S., 32, 61–62, 89
Suppes, P., 1, 44, 45, 94, 95
symmetry, 19, 35–45
    group of, 36
Symmetry Meaningfulness
    axiom of, 131
synthetic geometry, 40

## T

threshold function, 48
threshold measurement, 47–59
    Weber's Law, 52–55
Torgerson's Conjecture, 77–81
Torgerson, W. S., 77
total ordering, 21
transformation group, 36
    of a structure, 36
transitive, 21
translation by $r$, 49
translation scale, 28
Tversky, A., 1, 89

## U

union set, 106
uniqueness theorem, 22
unordered pair, 107

## V

Veit, C. T., 81

## W

Weber constant, 52
Weber representing structure, 52
Weber's Law, 52–56
    meaningfulness, 57–59
    modified Weber constant, 52
    qualitative characterization, 54
Weber, E., 53

## Z

Zimmer, K., 68
Zinnes, J., 45, 95